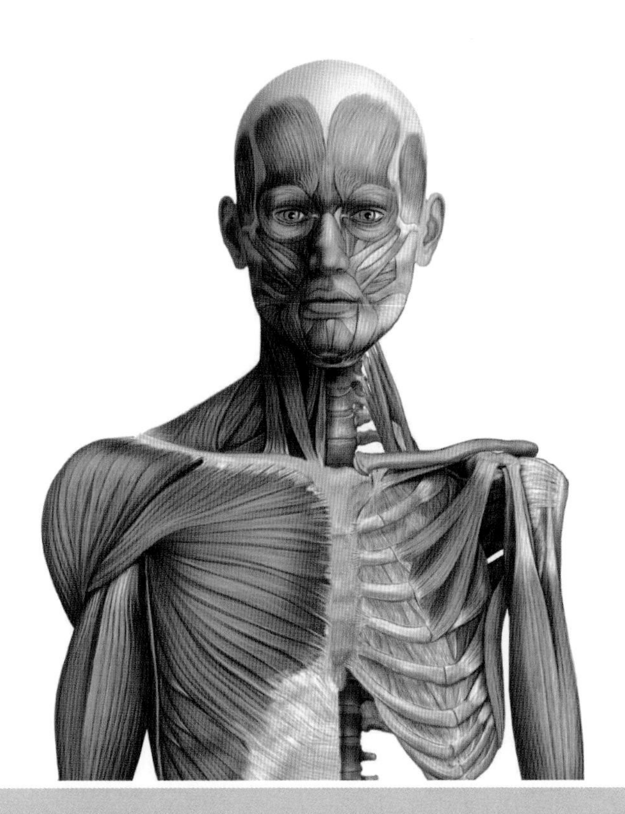

Anatomy & BODYBUILDING

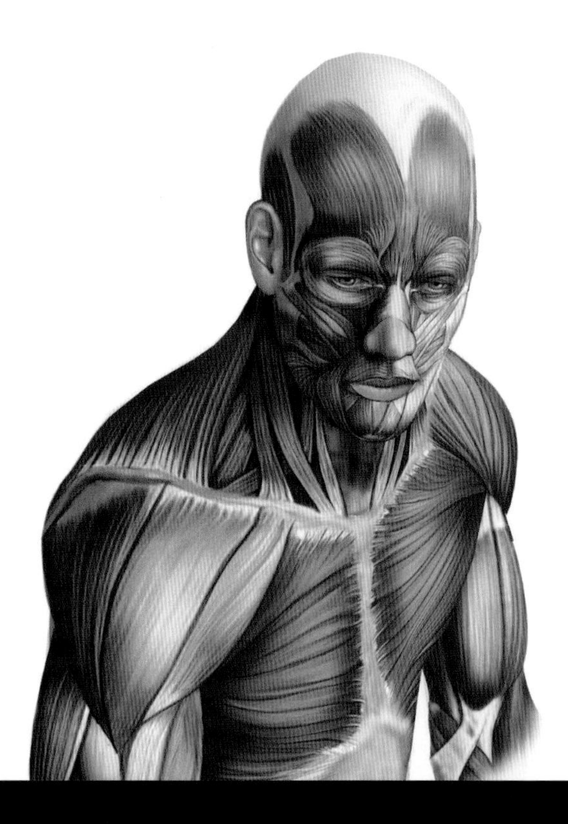

Anatomy and BODYBUILDING

First edition for the United States, its territories and dependencies, and Canada published in 2015 by Barron's Educational Series, Inc.

English-language translation © copyright 2014 by Barron's Educational Series, Inc.
English translation by Eric A. Bye, M.A.

Original Spanish title: *Anatomía & Musculación*
© Copyright 2014 by ParramónPaidotribo, S.L.—World Rights
Published by ParramónPaidotribo, S.L., Badalona, Spain

All inquiries should be addressed to:
Barron's Educational Series, Inc.
250 Wireless Boulevard
Hauppauge, NY 11788
www.barronseduc.com

ISBN: 978-1-4380-0548-5

Library of Congress Control Number: 2014930086

Production: Paidotribo Publishing
Editorial Mangement: Ángeles Tomé
Author: Dr. Ricardo Cánovas Linares
Editor: Guillermo Selijas
Technical Proofreader: Ana Lorenzo
Text Proofreader: Roser Pérez
Graphic Design: Toni Inglès
Illustrations: Myriam Ferron
Photographs: Nos & Soto
Layout: Estudi Toni Inglès
Pre-printing: Estudi Genis

Printed in China
9 8 7 6 5 4 3 2 1

Acknowledgments

I wish to express my gratitude to Paidotribo Publishing, especially to Emilio Ortega, for counting on me to produce this book, to María Fernanda Canal, for her wise and continuous assessments of the text, to Ángeles Tomé for her remarks and contributions and for her patience in our many meetings, to Guillermo Seijas, a great expert in physical conditioning, for his technical notes, which were extremely helpful, to Victor Cánovas, for his invaluable collaboration, and to everyone else who collaborated in producing this book.

To all, my most profound gratitude for the excitement, enthusiasm, and professionalism that they demonstrated. Thanks to them, I succeeded in this project so that people getting started in the world of physical activity and those who are already advanced in physical exercises can consider new horizons and achieve their goals.

The Protagonist Muscles

- **Agonist:** a muscle that works (contracts)
- **Synergist:** a muscle that contributes to the action of the agonist
- **Antagonist:** a muscle that performs the function opposite that of the agonist
- **Anchor Point:** support point for the movement

Identification of the exercise

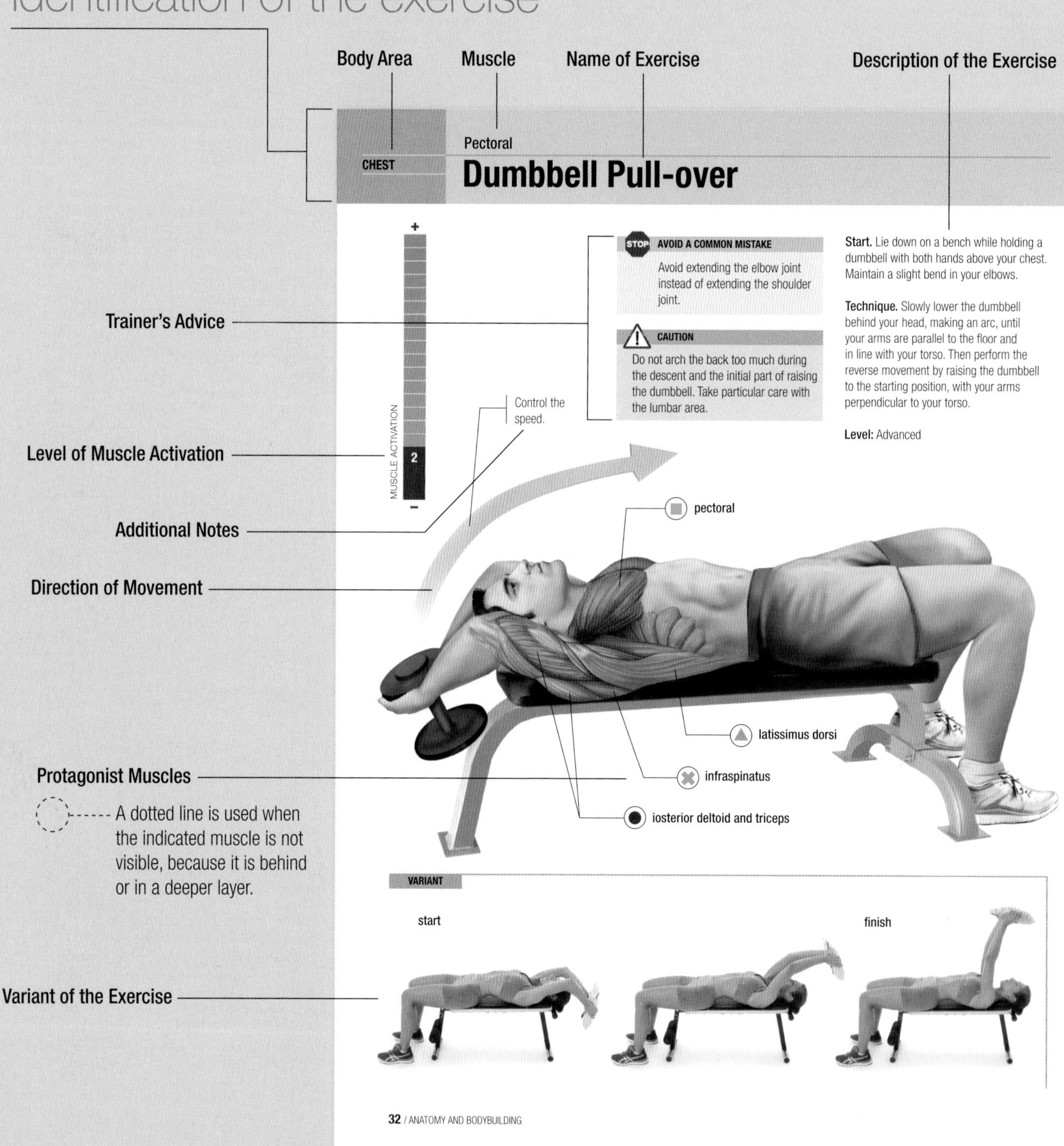

Body Area
Muscle
Name of Exercise
Description of the Exercise
CHEST
Pectoral
Dumbbell Pull-over
MUSCLE ACTIVATION
2
AVOID A COMMON MISTAKE
Avoid extending the elbow joint instead of extending the shoulder joint.
CAUTION
Do not arch the back too much during the descent and the initial part of raising the dumbbell. Take particular care with the lumbar area.
Start. Lie down on a bench while holding a dumbbell with both hands above your chest. Maintain a slight bend in your elbows.
Technique. Slowly lower the dumbbell behind your head, making an arc, until your arms are parallel to the floor and in line with your torso. Then perform the reverse movement by raising the dumbbell to the starting position, with your arms perpendicular to your torso.
Level: Advanced
Trainer's Advice
Level of Muscle Activation
Control the speed.
Additional Notes
Direction of Movement
pectoral
latissimus dorsi
infraspinatus
iosterior deltoid and triceps
Protagonist Muscles
A dotted line is used when the indicated muscle is not visible, because it is behind or in a deeper layer.
VARIANT
start
finish
Variant of the Exercise
32 / ANATOMY AND BODYBUILDING

Avoid a common mistake: warning to avoid mistakes that are frequently made

Caution: preventive notes

Notes from the Trainer

Performing the Exercise

Pectoral

Bench Press on the Machine

CHEST

Start. Sit at the pectoral press machine, with your shoulder blades pressing against the backrest and your arms parallel to the floor.

Technique. Raise both arms together, slowly and evenly, toward the front by straightening your elbows. Your elbows remain just beneath shoulder level. When you reach total extension of the elbows, avoid locking the joint. After a brief pause, slowly return to the starting point.

Level. Beginner, intermediate, and advanced

AVOID A COMMON MISTAKE

Keep your shoulder blades pressed against the back cushion.

CAUTION

Avoid pressing against the back cushion with your head and exerting pressure with your neck.

ANATOMY AND BODYBUILDING / **33**

Variant of the Exercise

In each exercise, we suggest a variant that you can do at home or away from the gym. All these alternative exercises can be done with low-cost equipment that is easy to find in sporting goods stores.

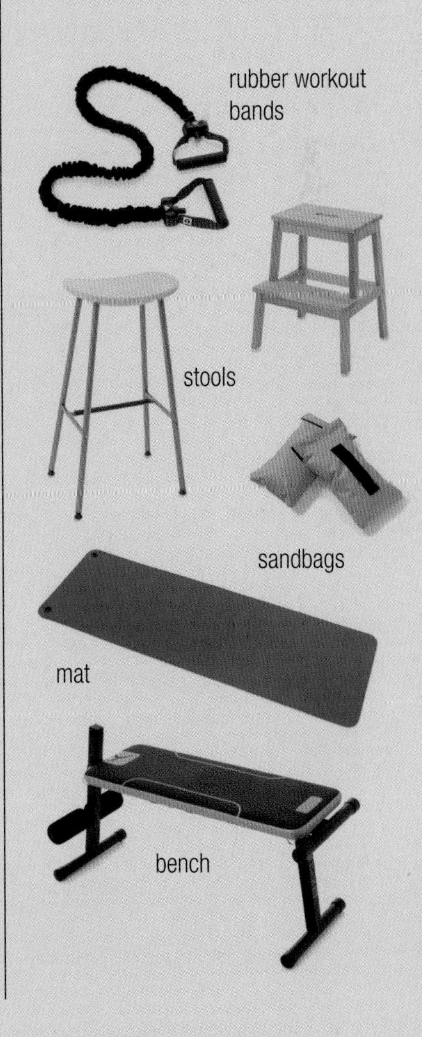

Before Beginning

Muscles and Training

Strength comes from force produced by muscles. It is possible to increase muscle strength significantly with a minimal investment of time. This is one of the principles of the method that is proposed in this book—training that is based on the principles of exercise physiology. This method consists of maximizing the muscle potential as far as your genetic makeup allows. For this purpose, intensity is vitally important. In order to maintain that intensity, muscles make use of a growth mechanism, but growth is not the same as size; otherwise, marathon runners would be extremely muscular. In other words, in order to increase its size, a muscle needs stimuli of high intensity, but fairly short duration.

It is possible to increase strength with a minimal investment of time.

Respect your muscle's language

This book is not about lifting weights, but rather about respecting the language of your muscles. The instant you stop maintaining muscle tension, your efficiency and, especially, your effort is in danger of being lost. Such a loss can then only be reversed by a patient effort adhered to conscientiously by you—in other words, by following a strict schedule of exercise.

This book does not suggest doing sets, because they are not cumulative. Let's take an example: when you drive a nail into place, there is no reason to continue hammering, because the only thing you would accomplish is to damage the surrounding area. Likewise, you shouldn't overdo your exercise routine to the point where you overwork your muscles. Beyond a certain point, additional exercise can be harmful to your muscles.

Our bodies have a limited capacity for dealing with the demands of stress, which is what exercise is all about. There are some guideline symptoms of stress, but they do not include stiffness. Stiffness indicates only incomplete cell metabolism. It does not indicate that you have worked correctly to improve your muscles.

Quite the contrary. If you try to play tennis for the first time, the next day you will experience major stiffness, and your muscles really will not have increased in size, not by a long shot.

The training that this book suggests involves muscle tension, working the muscles to the point of muscle failure, proceeding very slowly. This training is intended to help you build up your muscles without causing damage to your body.

Muscle mass

Muscle mass starts to decline at age 25–30. This reduction occurs independently of a person's activity level. But some studies have demonstrated that physical activity can keep us from losing muscle tissue if training focused on strengthening the muscle mass is carried out regularly and properly.

Such training makes it possible to continue to engage in activities that you used to engage in when you were younger. The strength reduction that occurs with age, in other words, is not inevitable. Strength is a function of the muscles, and it is possible to make great strength gains with a minimal investment of time.

This book shows you how to make these gains in an efficient and time-saving way.

The strength reduction that takes place with age is not an inevitable fact.

What is bodybuilding for?

Aside from the fact that it allows us to engage in physical activity (for amusement or competition), bodybuilding contributes significantly to health. It acts like a defense for our bodies for our whole lives.

All parts of the body age, but the only ones that can be rejuvenated are the muscles, and keeping them in good condition means that the other body parts will work better as well. For example, the liver will experience improved function in the presence of a proper muscle base. The same thing happens with the lungs and the heart, organs that in turn are used in improving muscle mass.

One of the many advantages of keeping the muscle system in good condition, aside from your being much stronger, is that it improves the cardiovascular system, which helps you to lose excess body fat, improves your endurance and flexibility, and increases your bone density. All this is possible through physical exercise carried out properly. And this is what is offered in this book.

Other advantages that muscles have is that they adapt to every demand placed on them. Behind every load of work done properly, there is rapid physical conditioning without injuries. The purpose of physical activity is to make you stronger and better prepared to engage in athletic contests or simply activities of daily life.

Bodybuilding, therefore, isn't simply about getting a more sculpted physique; it's also about improving your overall health and well being. It can significantly improve the quality of your life, and, possibly, the length of your life, if you put in the time and effort to let it have that effect.

How do muscles function?

The function of muscles is to contract—in other words, to bring their two ends closer together. In so doing, a muscle moves the bones closer together. It is strange that a muscle cannot extend itself. This is done through the contraction of its antagonistic muscle or through external pressure. The muscle goes from one extreme to another. The **origin** is the proximal area, which usually occupies a broad area of the bone. The **insertion** commonly is distal and sinewy, and it occupies a small area of the bone. Thus, the origin refers to a point where the muscle attaches to a bone and is normally less mobile.

Depending on the capacity for contraction, there are four types of muscle fibers, but to simplify we will focus on two: fast or white fibers and slow or red fibers.

Fast-contraction fibers are activated in situations that require explosive force of short duration, and slow fibers act in endurance exercises.

When we do an exercise, the slow-contraction fibers respond first. When the weight increases, those fibers begin to fail, and the intermediate and the fast-contraction fibers come into action.

The fibers exist in proportions already established at birth, and they vary from muscle to muscle and from person to person.

In high-intensity training, the purpose is to strengthen all types of muscle fibers, keeping in mind that you can work only with the genetic makeup you inherited: with age, as you grow older, you experience a loss of agility, speed, and so forth. Your fast-contraction muscle fibers begin to decay, and this makes it harder for you to carry out activities that used to be easy.

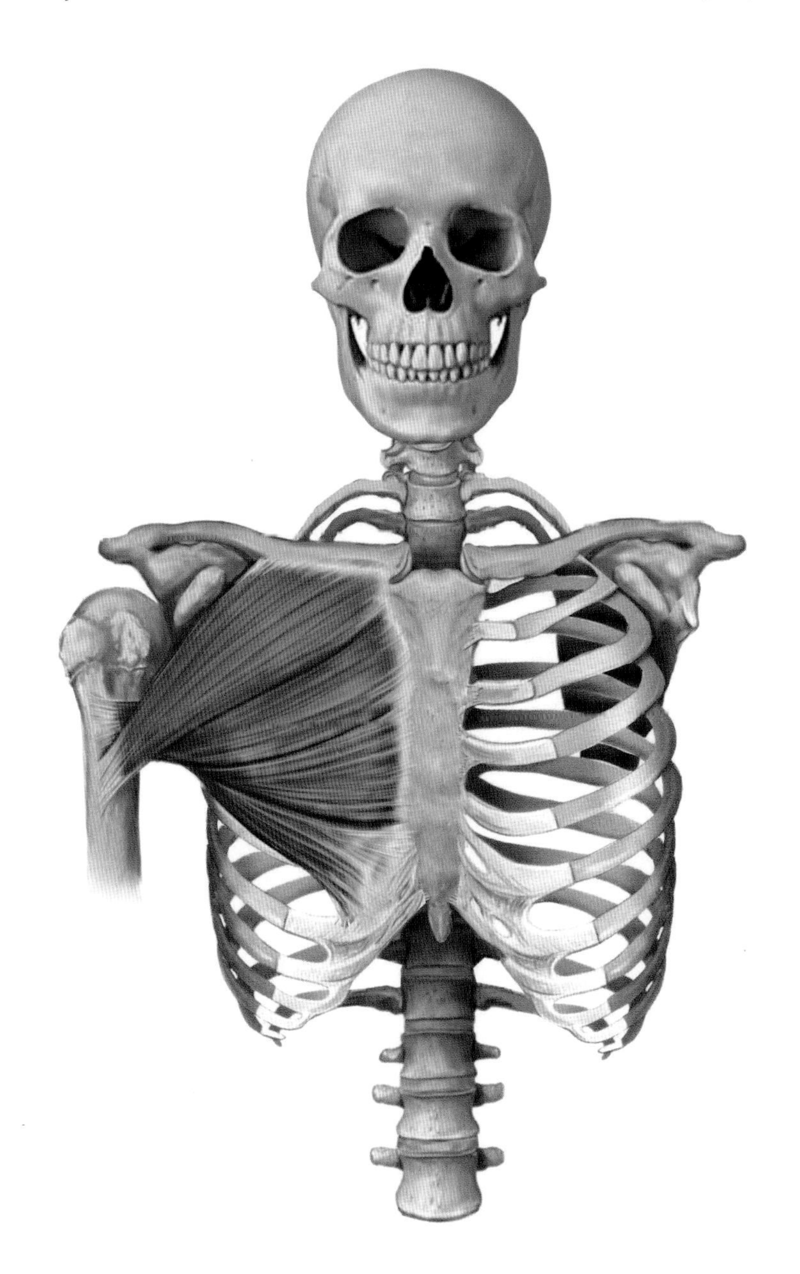

The proposed training: working until muscle failure

First of all, your training should be slow, but it will be much more intense than traditional training (repetitions of between one and two seconds). This makes it possible to burn up many more calories and more body fat, and it increases sensitivity to insulin (insulin resistance encourages obesity), so it regulates blood pressure, cholesterol, and triglyceride levels, as long as it is combined with a proper diet. And the most important thing: with a workout of less than a half-hour per week, you could realize a significant loss of body fat. By following this training properly, no muscle mass will be lost.

If you take in fewer calories than your body needs, aside from losing fat, you will lose also muscle mass as you exercise. If you add aerobic training to the mix, you will lose even more muscle. Thus, this type of training is destined to fail, because it involves a loss of muscle mass. What this book proposes is that you achieve a metabolism that allows you to burn fat, not muscle, due to the changes that high-intensity training involves.

In other words, to achieve muscle mass gains, you should not only follow the training regimen laid out in this book, but also adjust your eating habits and other exercise activities to realize this goal.

All parts of the body age, but the only ones that can be rejuvenated are the muscles.

By working the muscles to the point of muscle failure, you create a need in your body, because you have crossed over its normal threshold. It's as if the muscles are saying to your body: "There is a lot of work happening here."

Intense exercise stimulates an enzyme known as AMP kinase, which regulates some metabolic processes in your body. This enzyme becomes active during exercise, if you suffer from type-2 diabetes, or if you are obese—in other words, it regulates metabolism under conditions of abnormality.

Some studies have shown that this enzyme remains active for seven to ten days after a high-intensity workout. This explains how fat can be burned during the intervals between workouts—in other words, when you are at rest.

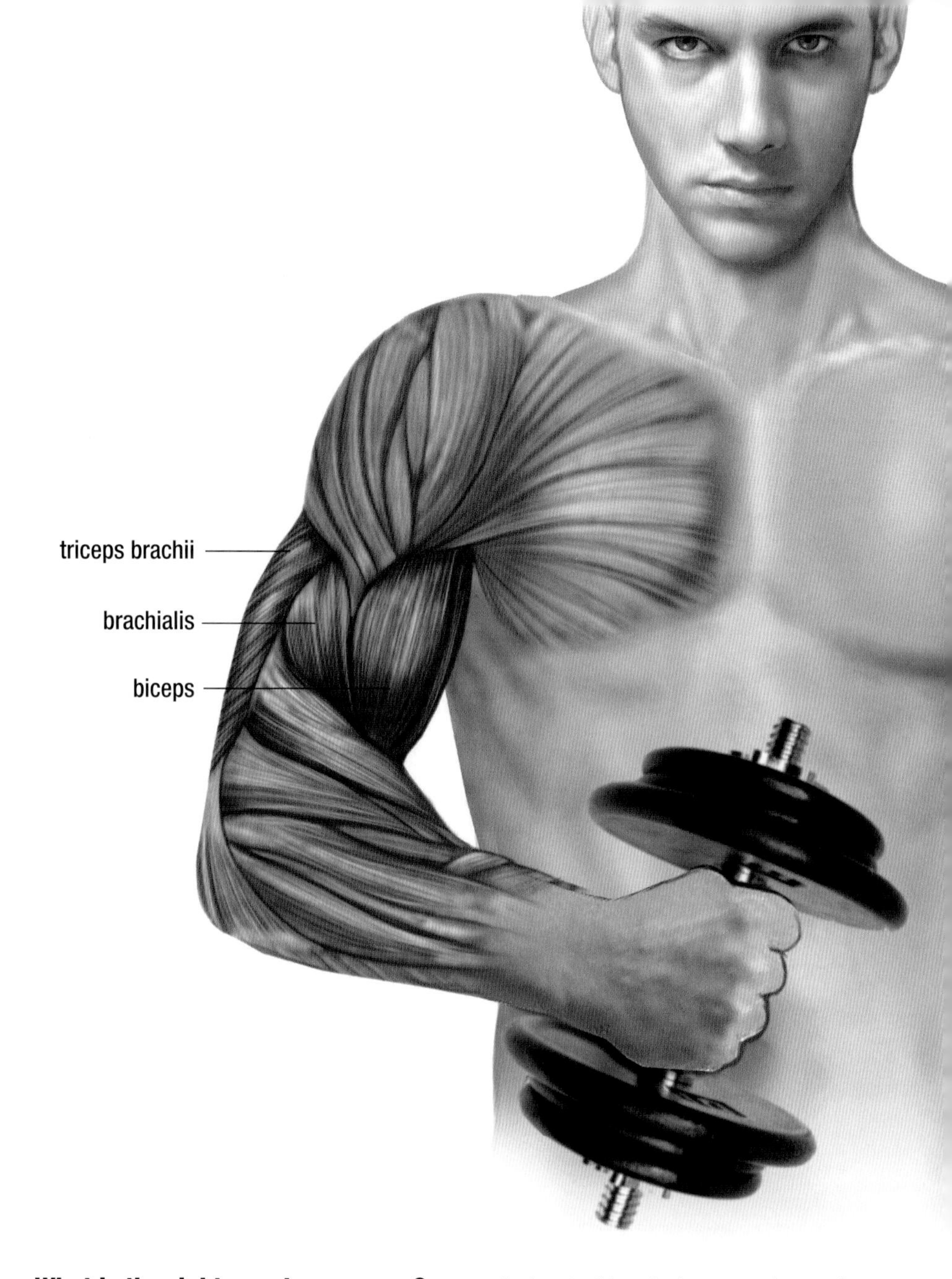

Agonistic, synergistic, and fixator muscles

In muscle movement, it is important to emphasize the role that each muscle plays. Thus, an **agonistic muscle** is the one that directly produces a given movement. **Antagonistic muscles** are the ones that act in opposition to the force and the movement that another muscle (the agonistic one) produces.

Another role that a muscle can assume is a **synergistic** one. Synergistic muscles act in conjunction with the **agonistic** muscles in performing movements. Finally, some muscles can act like a fixator or stabilizer, a very important mission that allows proper execution of muscle work; these muscles perform this fixating by means of isometric contraction. What a muscle can do does not indicate what it is going to do, so everything depends on stability.

What is the right way to warm up?

In your case, because the work is slow, you do not need a warm-up. The first 10–20 seconds of an exercise can be considered a warm-up. During this time, most of the active fibers are involved, and the inactive fibers are recruited in the last few seconds of the exercise, when all the fibers that have worked up to this point become exhausted. This is precisely what makes the workout effective.

How do we do a repetition?

A **repetition** is the action of repeating a specific exercise, and a **set** is a whole series of repetitions of that exercise.

In performing one repetition correctly, what you would consider impeccably, you must perform a slow, controlled movement that allows you to reduce the impulse to a minimum and increase the muscle tension. This way you will be doing high-quality repetitions.

It is also important to do a repetition properly in order to avoid injury. Don't overstress your body as you work out.

Thus, the most important thing is not how much weight you move, but rather how

It is crucial to be very conscious of your body position, especially the muscle fixation, and not to forget that you need to provide constant muscle tension.

you move it. This book recommends that the impulse be reduced to the minimum in the repetition, and that there be a brief pause at the maximum muscle contraction, if the exercise uses just one axis, and a very slow change of direction during the transition from the negative to the positive phase.

It is crucial to be very conscious of your body position, especially the muscle fixation, and not to forget the importance of providing constant muscle tension.

If weight is moved too quickly, the impulse will reduce the load on the muscle, making the exercise easier in the majority of the range of movement, and, of course, more dangerous. As a result, this book emphasizes the importance of lifting weight at a speed that forces the muscle to do as much work as possible. This usually happens through a cadence of four seconds of concentric work (when the muscle contracts, or positive work), two seconds of static contraction (when you hold the position without moving, or isometric work), and four seconds of eccentric contraction (when the muscle stretches out, or negative work). If weight is able to move quickly, there is insufficient stimulus for the muscle. Quickly lowering a weight from on high with a throwing motion, and the action of throwing, is not very useful for increasing strength.

This book advises emphasizing the eccentric phase of the repetition (when the muscle stretches out). It teaches how to prevent the acceleration of weight in the negative or eccentric phase. You must not let weight fall, because that will not contribute to the development of the muscle's size or strength.

Using the leg extension, for example, weight will have to be lifted slowly (concentric, or positive phase) and gently, at a speed at which the quadriceps has to work during the entire movement (about four seconds), in the extended position (when the legs are extended there should be a brief pause). Then you have to let weight down slowly (eccentric, or negative phase) over about four seconds. If you are not sure about the speed, you should raise and lower weight more slowly, not more quickly.

You must be conscious of the position of your body and muscle fixation. These two factors are important in exercising the muscle correctly and efficiently. All these details make the repetition easier. Sudden changes in the positions of your body make the exercise dangerous.

The ultimate purpose of training is to cause muscle tension, and this fact separates the experts from the beginners. You should understand the muscle as a machine that creates tension. To achieve this, it adapts to the force it must produce; thus, as you increase the demand, it becomes stronger.

The first repetition in a set is the most important one. Then it's the second one, which must be done in the same way as the first one. The goal is to reproduce perfect repetitions. If a set of repetitions were recorded on video, there should be no pronounced differences among them.

How do you determine how much to train?

When a person is asked how long a high-intensity workout should last, the answer of less than 30 minutes, once a week, is commonly met with a smile. Not everyone manages time in the same way, and intensity is equal to work divided by time.

This book considers the precise amount of training to be very important—no more, no less. Too much or too little may not produce the desired effect. (Of course, other books may offer different formulas, and this book doesn't claim that it offers the only formula for success.)

If you turn to an example from medicine, you will see things more clearly. If you are instructed to take a dose of a drug once every 24 hours, why shouldn't you take it every six hours? Simply because the therapeutic effects would not improve and the side effects would be very obvious—thus, the great importance of the dose. This example shows that a little bit stimulates and a lot inhibits.

In other words, too much exercise is just as bad as too little in regard to building up muscle mass. And too much exercise carries the additional risk of injury.

If the weight can be moved quickly, it is not adequate for stimulating the muscle.

How should we do the exercises?

The most important thing is to do them with the least amount of impulse. Muscles commonly respond if the set is done over 60–90 seconds. Over 90 seconds, the weight you are using must be increased by 5%; between 60–90 seconds, you should not vary the weight; and, under 60 seconds, you should take off 5% of the weight.

For example, if the weight you used in the previous workout was 155 pounds (70 kg) and the time was 96 seconds, next time you should increase the weight by 5%, by about 10 pounds (3.5 kg), so the new weight will be about 165 pounds (75 kg). If you are below the zone of 60–90 seconds, you should subtract 5% from the weight—in other words, you would work with a new weight of about 145 pounds (66 kg).

You should use four seconds to raise the weight and another four seconds to lower it; in some exercises, you will mark a brief pause between the two phases of the movement (in the technical explanation of each exercise, this book indicates whether it is necessary to include the pause, and, if so, for how long).

This way of working the muscle provides very significant muscle tension and preserves the joints (**progressive resistance**).

You must do repetitions until you can do no more—in other words, even though you can move the weight, you can't move it as far as in the previous repetitions (muscle failure). But don't overdo it to the point where you could injure yourself.

What's the intensity of the workout?

It is the maximum effort that is exerted in trying to arrive at point Y from point X. For example, in lifting a weight (biceps exercise), if you bend your elbow and lift the weight toward your shoulder, before raising the weight, it is at point X, and, when the arm is bent, it is at point Y. When the difficulty is such that, by bending the forearm onto the upper arm, you cannot reach point Y, than there is muscle failure. This effort is known as the intensity of the workout.

When you work in this manner, you need to do just one set of each exercise.

Your body will not do anything if it doesn't have a reason to do it, so reaching muscle failure—in other words, working at high intensity—will wake up the muscle fibers that never would have become involved in muscle contraction.

Progression

In order to achieve your purpose, you need to aim for a goal step by step. To do that, you must apply progressive resistance—in other words, you must try to add weight or time in every workout.

You must always be aware of your progress. For that purpose, you should use a training log.

It is important not to make the mistake of comparing yourself to others. The only person you can compare yourself to accurately is yourself. You cannot rate the success of your training program according to what other people are doing.

The key to continual progress consists of balancing three elements: the high intensity of the training, the progressive overload of the training, and the frequency of training.

Here is a training model for the entire body, in which about two minutes are dedicated to each exercise:

1. leg press
2. hamstring curl
3. calf exercises
4. pectoral press
5. cables for the back
6. shoulder press
7. biceps curl
8. french press
9. abdominal curl
10. lower back extension

Beginners

This book considers people who have no experience in weight training, as well as athletes who have not trained for years, to be beginners.

Such people must begin with a program in which the resistance is light and the short-term goal involves performing the exercise correctly. The progression of the training program must come over time, as the effort of the workout increases. Programs must be based on the abilities of individuals and their goals; these programs are centered mainly on increasing the load gradually. As the increase in muscle work approaches the limit of a person's genetic ability, so progress becomes more difficult.

Intermediates

People who are at an intermediate level have shown a clear commitment and regularity in their training. It is commonly said that, at this level, one has to give increasing attention

Y

X

What people involved in high-intensity workouts need most is brevity.

to the frequency of the training and the incorporation of a split routine in the training program. Nothing is further from the truth, however. Using this book's criteria, based both on experience and on scientific studies, intensity is at odds with the amount of work, so brevity is what people who are involved in high intensity need most.

How much time should people who do the intermediate workouts spend on the same exercise? Until they can continue no longer or lift any more weight. When the exercise can no longer be done, it must be replaced by another one that allows continued progress. Every exercise has a limit, and progress can stagnate by reaching the maximum (genetic) strength potential or by overtraining.

How much weight should be used?

Beginners

This book could suggest some calculations and formulas, but the quickest and most practical way to figure out the best weight for you is to handle a very light weight (one with which you can do 20 repetitions without great difficulty) and concentrate, during the first three workouts, on doing the exercises correctly. From then on, you can increase the load.

Intcrmediates

Starting with 12 sessions (3 months), high intensity can be applied until you reach muscle failure.

The key to constant progress involves balancing three elements: the high intensity of the training, the progressive overload of the training, and the frequency of training.

You stimulate your muscle mass in the gym, but it grows during rest.

Advanced

When you do exercises, you can reach plateaus (that is, you can't get beyond a certain weight in at least three consecutive workouts) that will keep you from moving ahead. High-intensity techniques can be applied to get over these plateaus.

Weight-reduction training (breakdown) is a productive procedure for getting over the plateaus. This is the most basic method for extending the exercise set. When you cannot continue with a set, you should reduce the weight by 15% and continue until you reach muscle failure (for the second time). Here, the work time for the set needs to be modified. So, from this point on, you should extend the set by reducing the weight as previously indicated (by 15%).

After 6 weeks of weight-reduction training (breakdown), you should experience significant improvement in muscle development. At this point, you need to return to standard training for an equivalent time (six weeks) to reduce the risk of overtraining, to save time, and to reach a high rate of productivity.

How can I increase my muscle mass?

Strangely, it's not in the gym that your muscle mass increases; there, you only stimulate it so that it grows, and this growth occurs only if we allow it to, by resting.

Stimulus → Recovery → Growth

Training log

To evaluate your progress, it is essential to keep a chart. This book recommends maintaining a training log, with which you can confirm, modify, or correct any outcome. This will allow you to evaluate your training with objective data.

The following is one example of a training log:

Name / First Name

	Weight / timo					
Date	May 21					
Weight	165 lbs. (75 kg)					
Skin Fold	.58 in. (15 mm)					
Press	70/110					
Leg Curl*	35/87					
Calves*	70/110					
Pectoral Press	65/92					
Machine Row	45/102					
Shoulder Press	50/85					
Latissimus dorsi cable	42/66					
Abdominals (specify)						
Lumbar (specify)						

Notes: * optional

The value of writing down the results of a workout so you can evaluate your progress from week to week can't be stressed enough.

One of the reasons why many people never succeed in changing their muscle tone is that they do not maintain a detailed plan of everything they do during a workout. They usually base everything on their memories, or they simply always train with the same weight, without having reached their genetic limits.

Should you change your workout?

Even if a muscle and its function do not change over time, this is not a reason to change an exercise. The key to improving a muscle's growth seems to be progressive resistance, not a change in routine, because this is not a question of distracting the muscle.

Body Composition

You should not take your cues from your ideal weight, but rather from the composition of your body—in other words, from the ratio between muscle mass and fat.

To estimate your body's fat deposits, you can measure the skin fold over the triceps on your upper arm using a fat caliper, with the aid of a friend, or measure your abdominal circumference with a measuring tape. To measure your muscle mass, you merely need to weigh yourself on a scale.

Before starting to work out, you will need to know your weight and check your skin fold. So, after ten sessions, when you take these measurements again, you will get reliable information and confirm that you have been training properly. Why ten sessions? Because your body does not change in linear form, but goes through ups and downs. To identify the tendency, therefore, you need a minimum of ten weeks.

How is the skin fold measured?

To measure the skin fold on your arm, you must measure the fold over your triceps. Your arm needs to be relaxed and extended along the side of the body.

The distance is measured between the acromion (the apophysis of the triangular shoulder blade, which joins the end of the clavicle) and the olecranon (one of the apophyses of the upper part of the ulna, which is shaped like a prism, with a quadrangular base, and which makes up the dorsal prominence of the elbow). The midpoint between the two is marked.

About 3/8 in. (1 cm) from the midpoint mark, you take a vertical pinch of skin and fatty tissue of the triceps (on the back of the upper arm). You must be sure to take only skin and fatty tissue, not muscle. If you are not sure, the muscle can be flexed. If muscle has been included in the pinch, you will see it tighten when flexing the muscle. In this case, you will have to release the skin and try again.

The caliper is applied to the skin fold about 3/8 in. (1 cm) from the fingers. There is no need to loosen the fingers during the measuring process. The caliper must exert constant pressure when taking the reading.

You need to write down the results of your workout you are doing so that you can evaluate your progress over time.

After three seconds, the reading is done to the nearest millimeter and the value is recorded.

Finally, the caliper is removed and the skin is released.

This process must be repeated three times, and the average of the three measurements is calculated. If one of them differs from the others by more than 10%, it must be discarded and a fourth measurement must be taken.

How is the abdominal circumference measured?

After ten sessions, when you take new measurements, the following may occur:

- Body weight has increased, but the skin fold measurement has gone down. This indicates an **increase in muscle mass**. You are working out correctly, and you are getting closer to your goal.

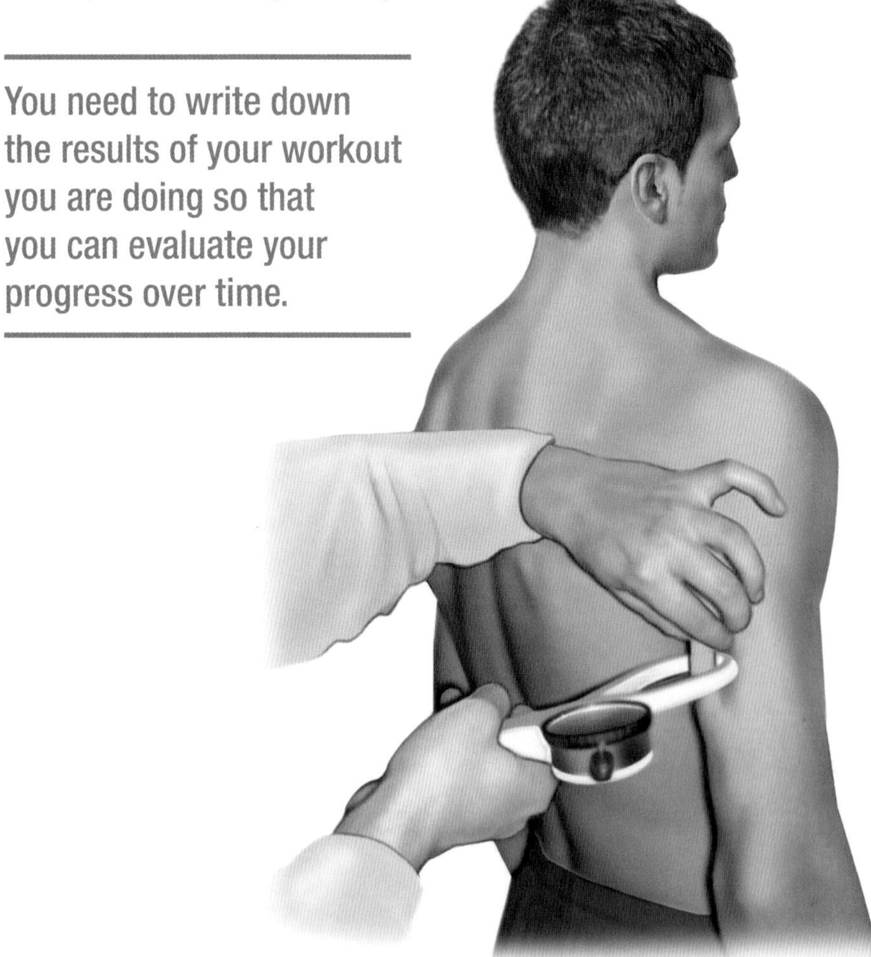

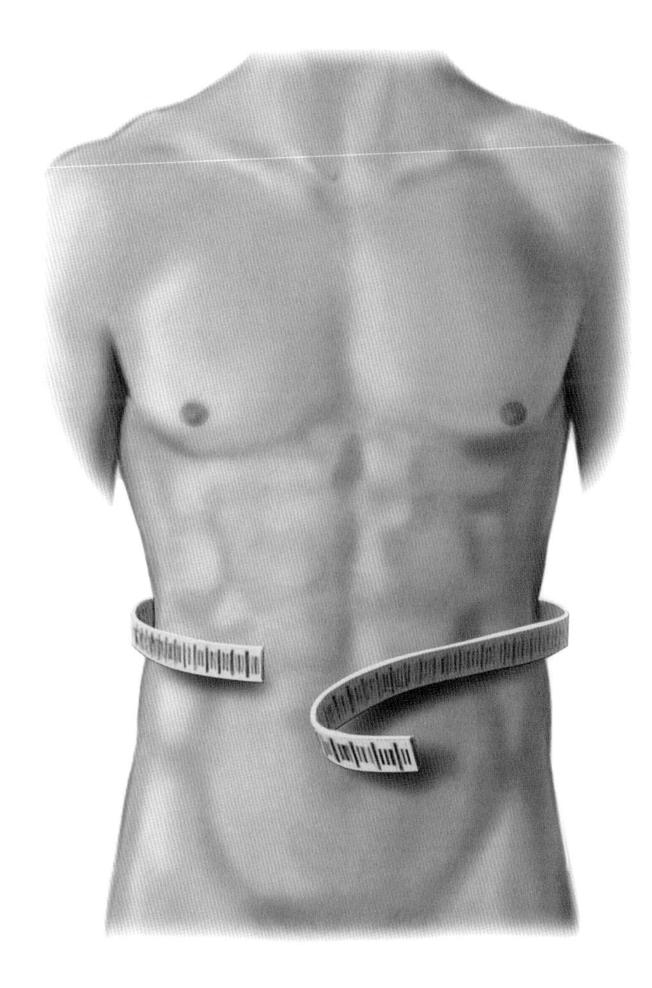

• Body weight goes down with no reduction in the skin fold measurement. This indicates a **loss of muscle mass**. You need to review your workout program:

– Are you sticking to the indicated workout time or are you overtraining?

– Are you sticking to the rest time?

– Are you performing the proper range of movement?

• Weight and the skin fold measurement both go down. This may indicate a **loss of fat**. If this is your goal, you are on the right track; otherwise, you will have to re-examine your workout program and your diet.

Muscle Activation

Electromyography (measuring electrical activity inside the muscles) is an interesting contribution to muscle activation, and it offers the possibility of measuring and comparing the electrical activity in one muscle group when exercising.

The muscle's electrical activity—and along with it, muscle tension—is significantly affected by biomechanical factors, such as the weight used and the behavior of the levers.

Analyses of the exercises using this system make it possible to compare various strength exercises for a specific muscle.

It is very important to point out that the surface electrodes record the activity of all the muscles located beneath the attachment point, which is useful in evaluating the work of the muscles.

This information helps you to understand the muscle you want to work with much greater clarity. Plus, you can distinguish whether the weak point is true or false.

A false weak point is weak because it has never been worked, whereas a true weak point is weak because genetics did not intend it for this purpose.

How do we use this information for our training?

From among the suggested exercises, you must select the ones that you focus on less often.

Before beginning to train a muscle or area, it is important to know how the muscles work in your own body. For this, the best course is to choose an exercise with more muscle activation than your routine indicates; this way, you educate yourself about the work that you must do in training to improve their performance.

Myths

I have bad genetics

People don't improve, and they blame their genetics. Many people who spend hours in the gymnasium with very poor results end up blaming their genetics rather than their training.

True, genetics exert a lot of influence on results. But much less than you might believe. It is clear that genetics is important; but it is not an obstacle to gaining muscle or losing fat. You merely need to train in the proper way to accomplish these things.

I can't add much muscle because I am thin

Some people believe that they can't add muscle mass, even if they train intensively or take in great quantities of food. It is a fact that some people succeed more easily in increasing their muscle mass than others. But this does not mean that those others are condemned to be saddled with the physique they now have.

You have to do lots of abdominal exercises to have sculpted abs

It's much easier to have impressive abdominals than you might think.

How do you get them? You must lose 10% of your body fat if you're a man and 15% if you're a women.

You can't burn fat from the abdomen by doing abdominal exercises, or with any other type of exercise. What you have to do to burn belly fat is to follow a proper diet and the training that this book suggests.

Atlas of the Muscular System

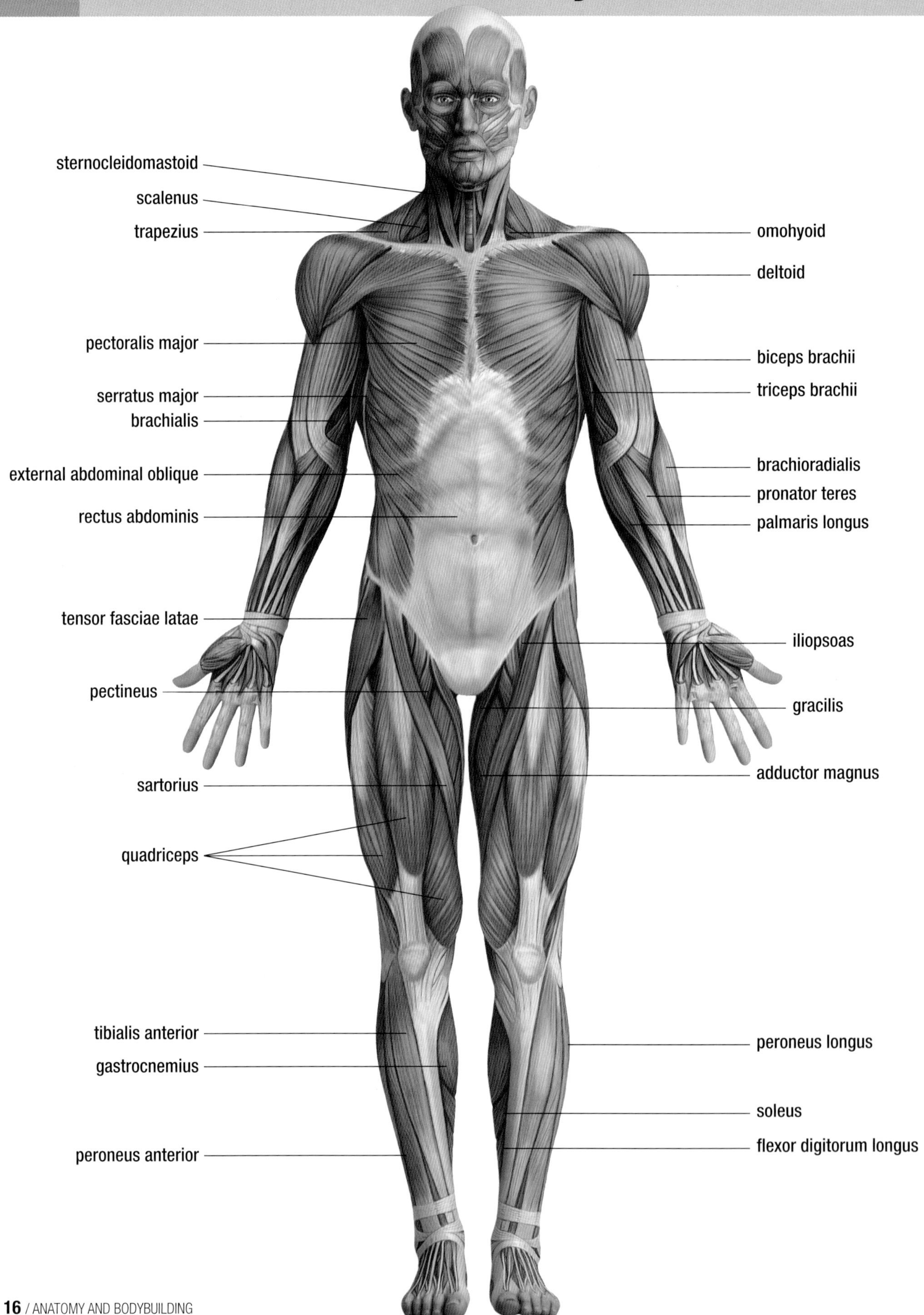

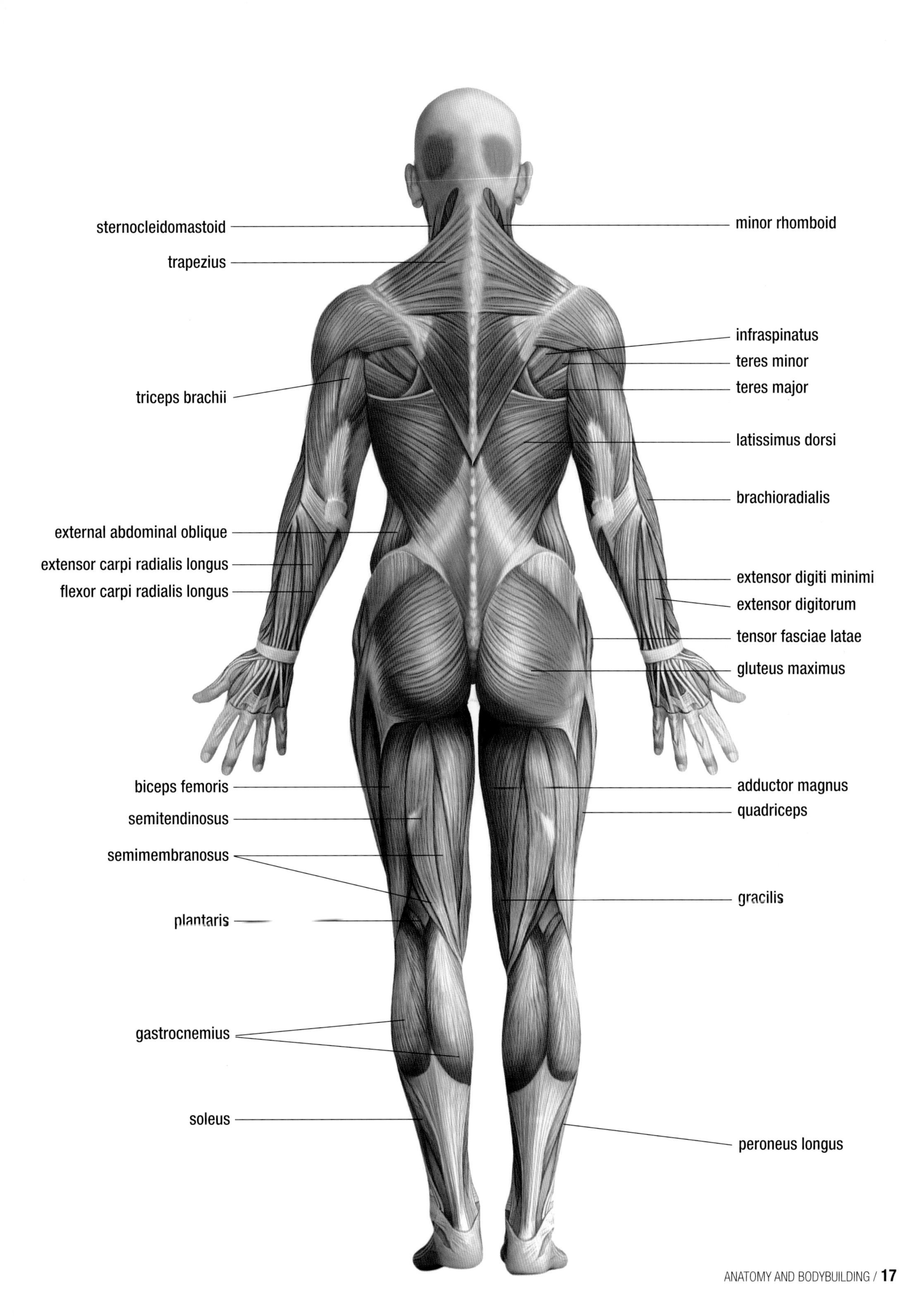
sternocleidomastoid
trapezius
minor rhomboid
infraspinatus
teres minor
teres major
triceps brachii
latissimus dorsi
brachioradialis
external abdominal oblique
extensor carpi radialis longus
flexor carpi radialis longus
extensor digiti minimi
extensor digitorum
tensor fasciae latae
gluteus maximus
biceps femoris
semitendinosus
semimembranosus
plantaris
adductor magnus
quadriceps
gracilis
gastrocnemius
soleus
peroneus longus

Planes of Movement

In order to understand the explanation of an exercise technique, it is necessary to have a clear idea of certain concepts that are commonly used to describe the movements of the human body. The first thing we need to know is that these movements are associated with three distinct planes: frontal, sagittal, and transverse. And, we have to consider that many movements are done in mixed planes.

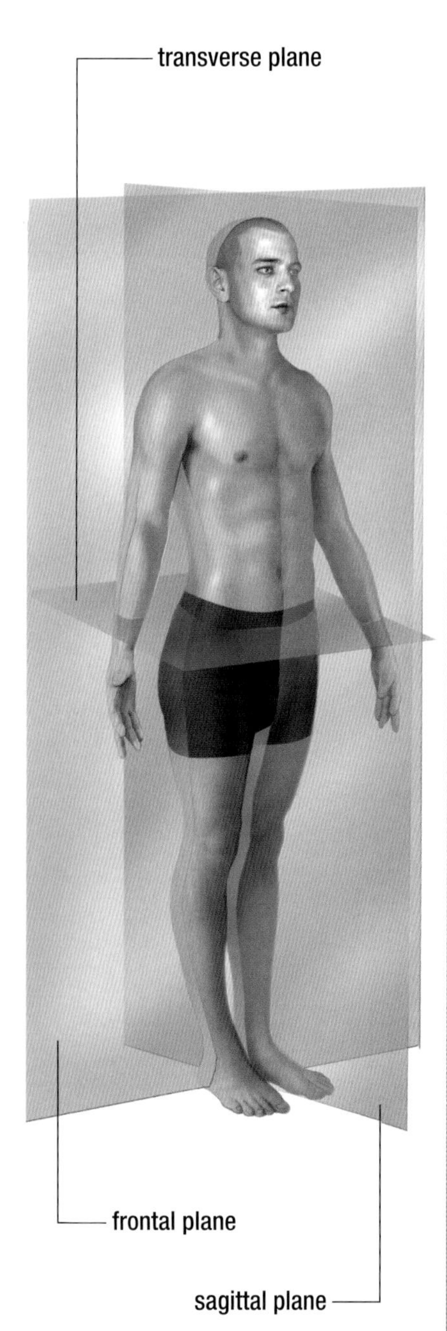

Frontal Plane

This is the plane that divides the body into frontal and posterior sections, and this is the plane where the movements of abduction, adduction, and lateral inclination take place.

Abduction. This is a movement in which we move a limb away from the central axis of the body. This distancing must be perceived from the front by the individual performing the movement. When you raise an arm to one side to hail a taxi, for example, you are using shoulder abduction.

Adduction. This kind of movement is the opposite of the preceding one. It results when one brings a limb in toward the central axis of the body. Once the taxi has seen you, for example, you lower your arm and place it next to you body, performing shoulder adduction.

Lateral Inclination. This is a movement in which an individual drops his or her head or torso to one side, as seen from the front. When you are sitting and nod off, your head commonly sags to one side, thereby producing a lateral flexion of the neck.

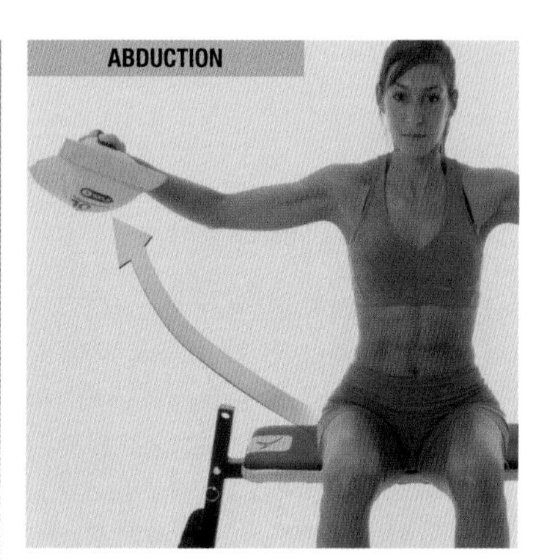

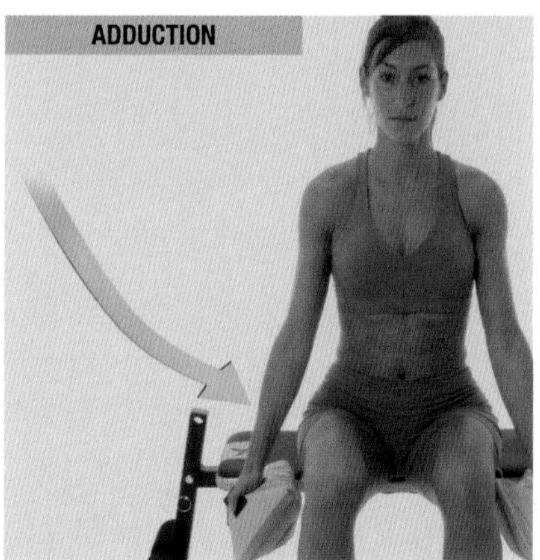

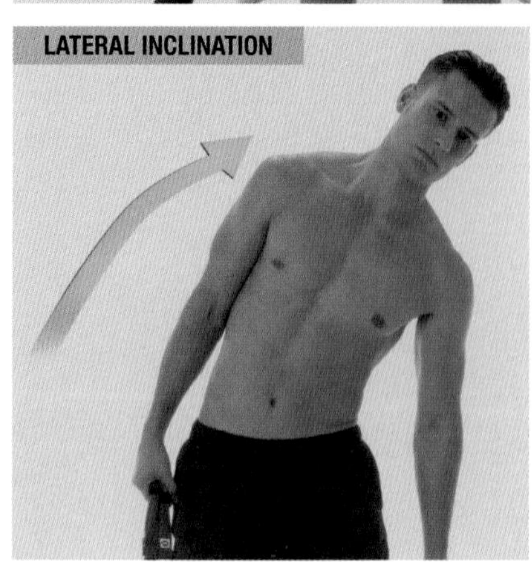

Sagittal Plane

This is the plane that divides the body into right and left sections. In this plane, you have to distinguish the movements of flexion, extension, antepulsion, and retropulsion.

Flexion. This is a movement in which you move a part of your body forward with respect to the central axis. There are

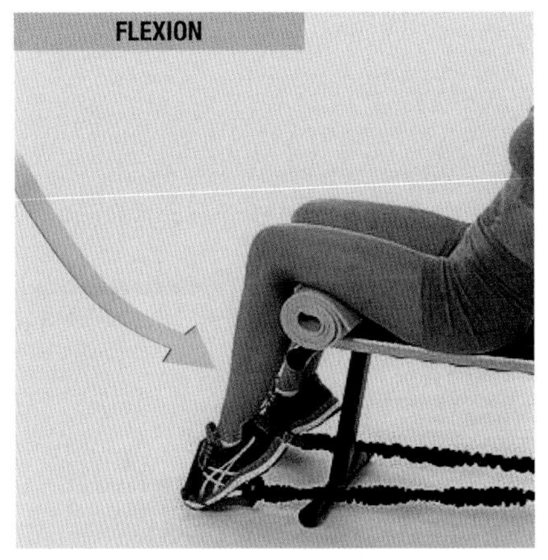
FLEXION

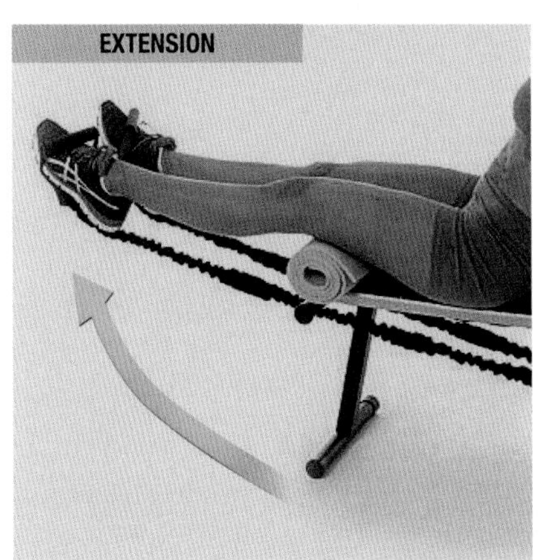
EXTENSION

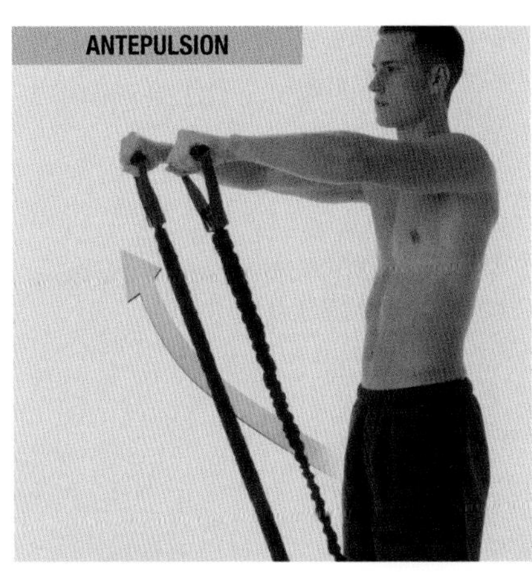
ANTEPULSION

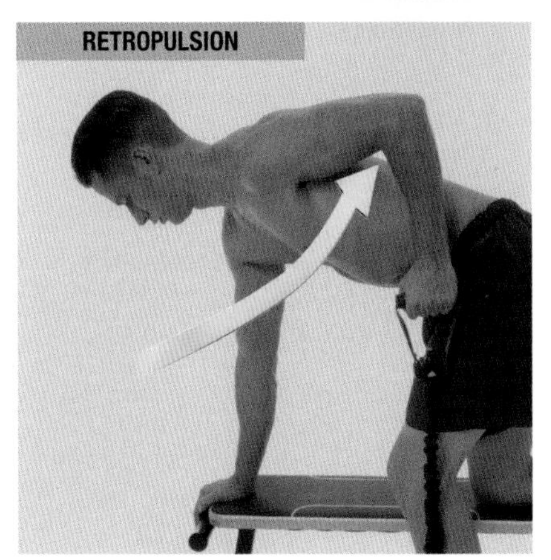
RETROPULSION

some exceptions, as in the case of the knee, where it is the opposite. This kind of movement is seen more effectively from the side of the person doing it, by looking at the individual in profile. If you are standing and you sit down in an armchair in such a way that your knees remain in front of your torso, you have performed hip flexion.

Extension. This is a movement in which you move a part of the body rearward with respect to the central axis. As in the previous case, this is the inverse when speaking of the knee. Let's consider the case of tying a necktie. When you finish the knot and relax your arms so that they hang alongside your body, you are performing an extension of the elbows.

Antepulsion. This is equivalent to flexion, but applicable exclusively to shoulder movement.

Retropulsion. This is equivalent to extension, but applicable exclusively to shoulder movement.

Transverse Plane

This is the plane that divides the body into the upper and lower sections. The transverse plane is where external and internal rotation, pronation, and supination take place.

Outward Rotation. This is a movement in which you move a part of your body away from the central axis of the transverse plane. If you are in a standing position and you decide to move the points of your feet outward penguin-fashion, you are performing an outward hip rotation. The foot will experience a turn on its own axis.

Inward Rotation. This is a movement in which you bring a part of your body closer to the central axis inside the transverse plane. If you get tired of standing like a penguin and you decide to move the tips of your feet inward—in other words, together—you are performing an inward hip rotation.

Pronation. This is a rotational movement of the forearm that allows you to put the back of your hand upward and the palm downward. When you grasp the handlebar of a bicycle, your hands are in pronation.

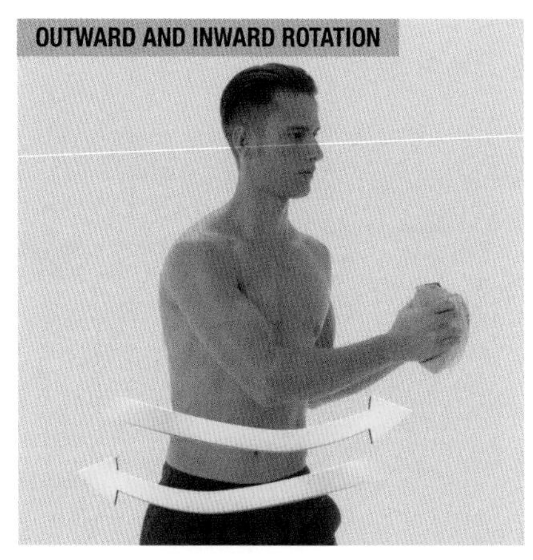
OUTWARD AND INWARD ROTATION

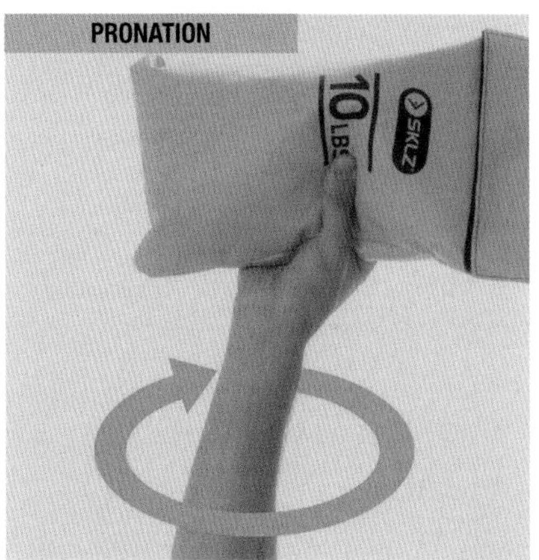
PRONATION

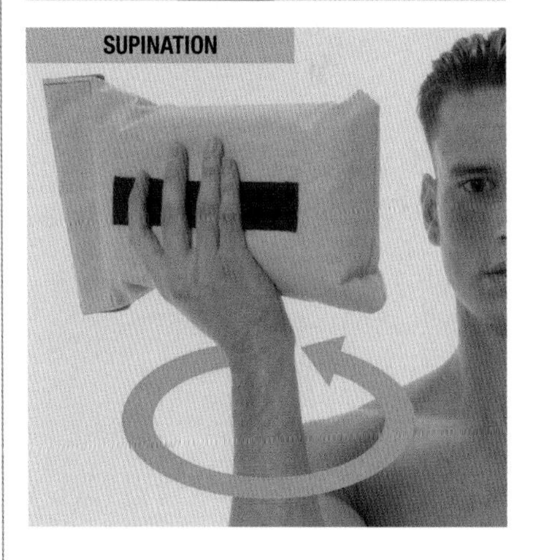
SUPINATION

Supination. This is a rotational movement of the forearm by means of which you can place the palms of your hands upward. If you pay for a purchase and you are handed change, you put your hand into supination, with the palm facing upward, to receive the coins or bills.

Chest

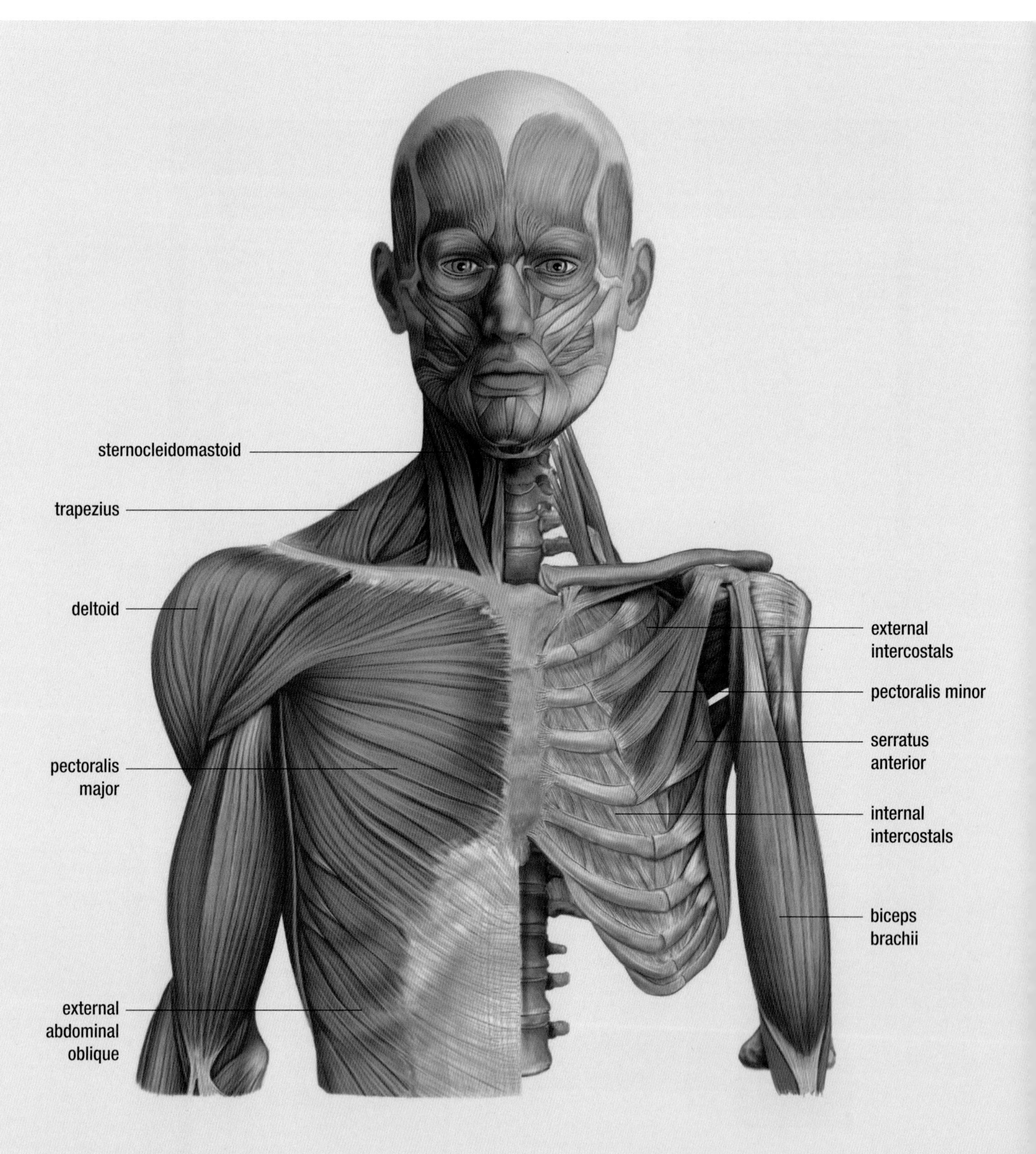
sternocleidomastoid
trapezius
deltoid
pectoralis major
external abdominal oblique
external intercostals
pectoralis minor
serratus anterior
internal intercostals
biceps brachii

This is the area in which the anterosuperior section of the torso is located. The largest and most powerful muscle in this area is the pectoralis major. Still, you must not forget that, even though you may not consider them to be important in bodybuilding, there are other muscles in the chest area, such as the pectoralis minor, the serratus anterior, and the intercostal muscles.

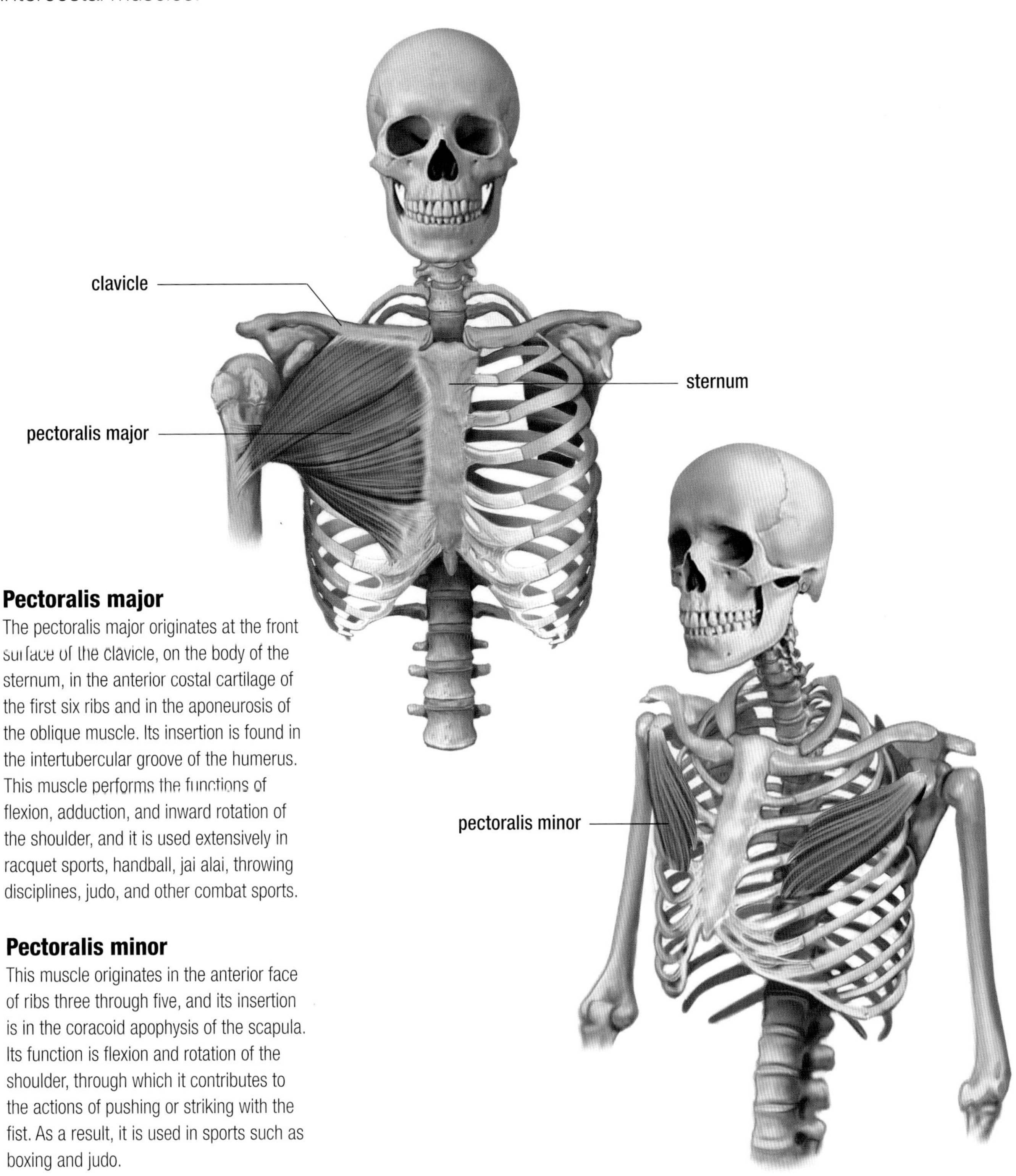

Pectoralis major

The pectoralis major originates at the front surface of the clavicle, on the body of the sternum, in the anterior costal cartilage of the first six ribs and in the aponeurosis of the oblique muscle. Its insertion is found in the intertubercular groove of the humerus. This muscle performs the functions of flexion, adduction, and inward rotation of the shoulder, and it is used extensively in racquet sports, handball, jai alai, throwing disciplines, judo, and other combat sports.

Pectoralis minor

This muscle originates in the anterior face of ribs three through five, and its insertion is in the coracoid apophysis of the scapula. Its function is flexion and rotation of the shoulder, through which it contributes to the actions of pushing or striking with the fist. As a result, it is used in sports such as boxing and judo.

Pectoral

Incline Press with Dumbbells

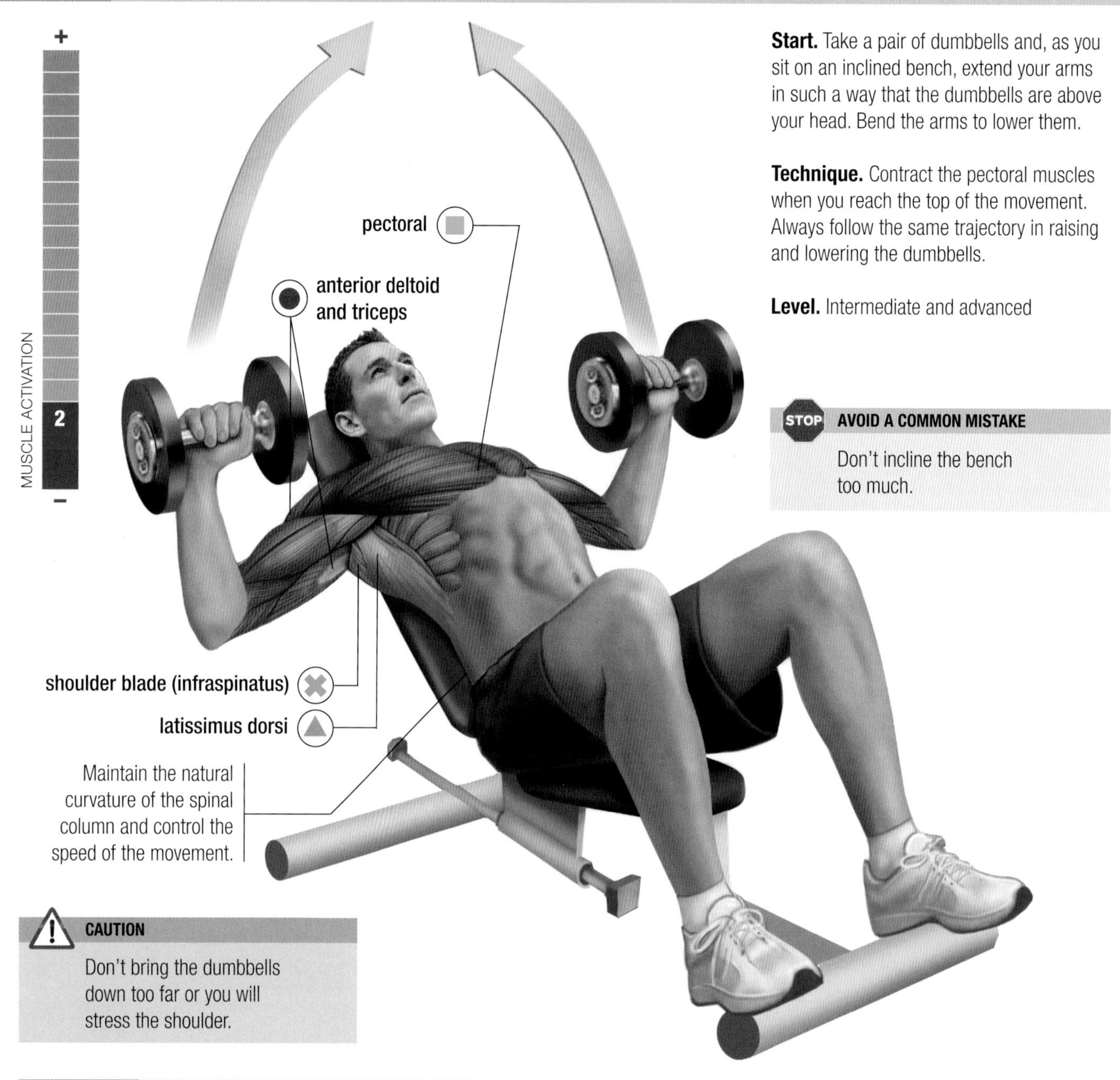

Start. Take a pair of dumbbells and, as you sit on an inclined bench, extend your arms in such a way that the dumbbells are above your head. Bend the arms to lower them.

Technique. Contract the pectoral muscles when you reach the top of the movement. Always follow the same trajectory in raising and lowering the dumbbells.

Level. Intermediate and advanced

STOP AVOID A COMMON MISTAKE

Don't incline the bench too much.

CAUTION

Don't bring the dumbbells down too far or you will stress the shoulder.

VARIANT

Cable Crossovers

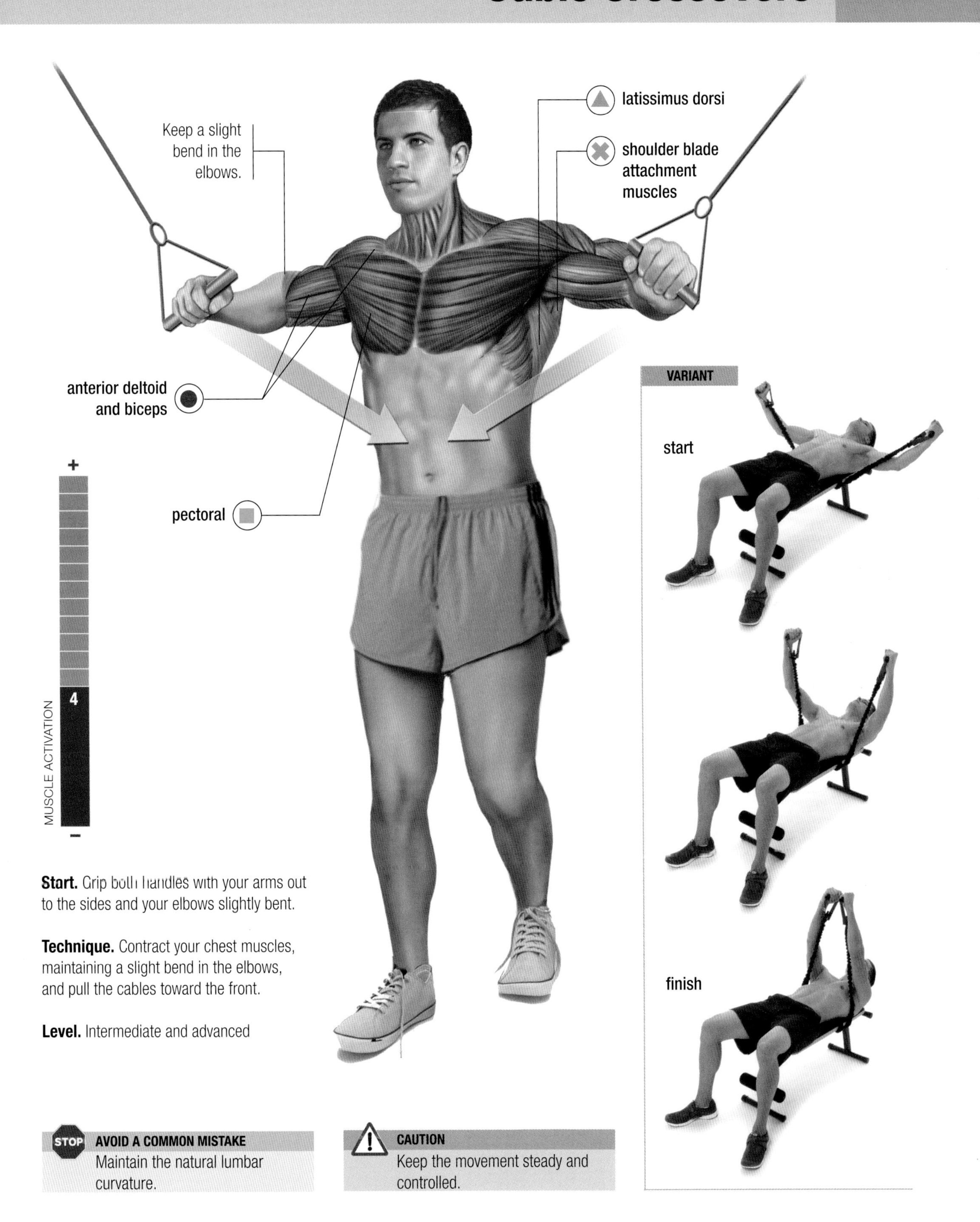

Start. Grip both handles with your arms out to the sides and your elbows slightly bent.

Technique. Contract your chest muscles, maintaining a slight bend in the elbows, and pull the cables toward the front.

Level. Intermediate and advanced

AVOID A COMMON MISTAKE
Maintain the natural lumbar curvature.

CAUTION
Keep the movement steady and controlled.

Pectoral

Bench Press with Dumbbells

Start. Take two dumbbells and lie down on the horizontal bench. Start from high, with your arms almost totally extended (but not locked).

Technique. Lower the dumbbells to the level of the pectoral. Pause briefly and go back up to the starting position.

Level. Beginner, intermediate, and advanced

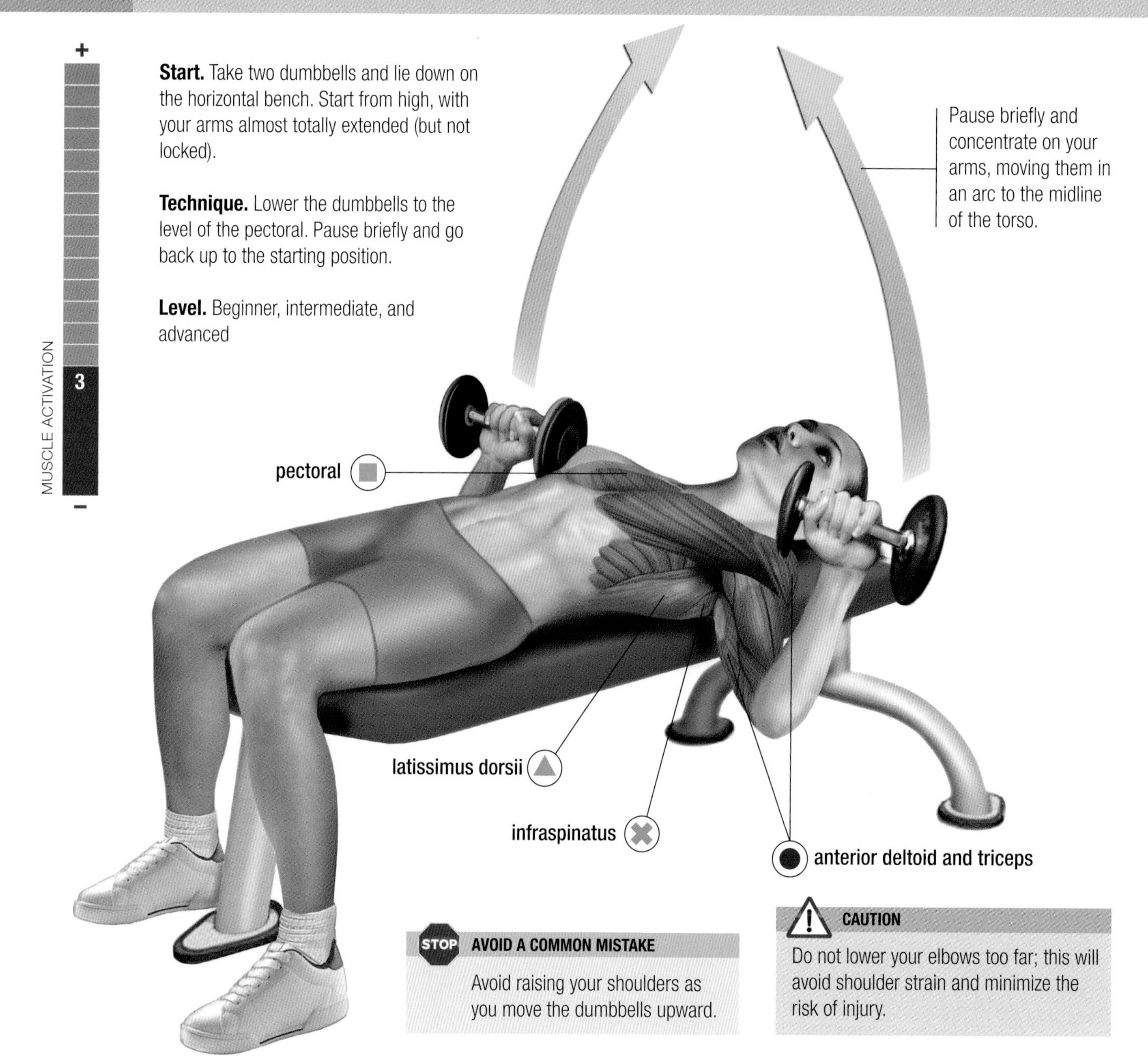
+
MUSCLE ACTIVATION
3
–
Pause briefly and concentrate on your arms, moving them in an arc to the midline of the torso.
pectoral
latissimus dorsii
infraspinatus
anterior deltoid and triceps

STOP AVOID A COMMON MISTAKE

Avoid raising your shoulders as you move the dumbbells upward.

CAUTION

Do not lower your elbows too far; this will avoid shoulder strain and minimize the risk of injury.

VARIANT

start
finish

Bench Press with Barbell

Start. Lie down on a level bench, take the bar with a grip slightly wider than your shoulders, and life the bar off the supports.

Techinque. Lower the bar to your chest, at which point your forearms will be perpendicular to the floor, and return to the starting position.

Level.
Intermediate and advanced

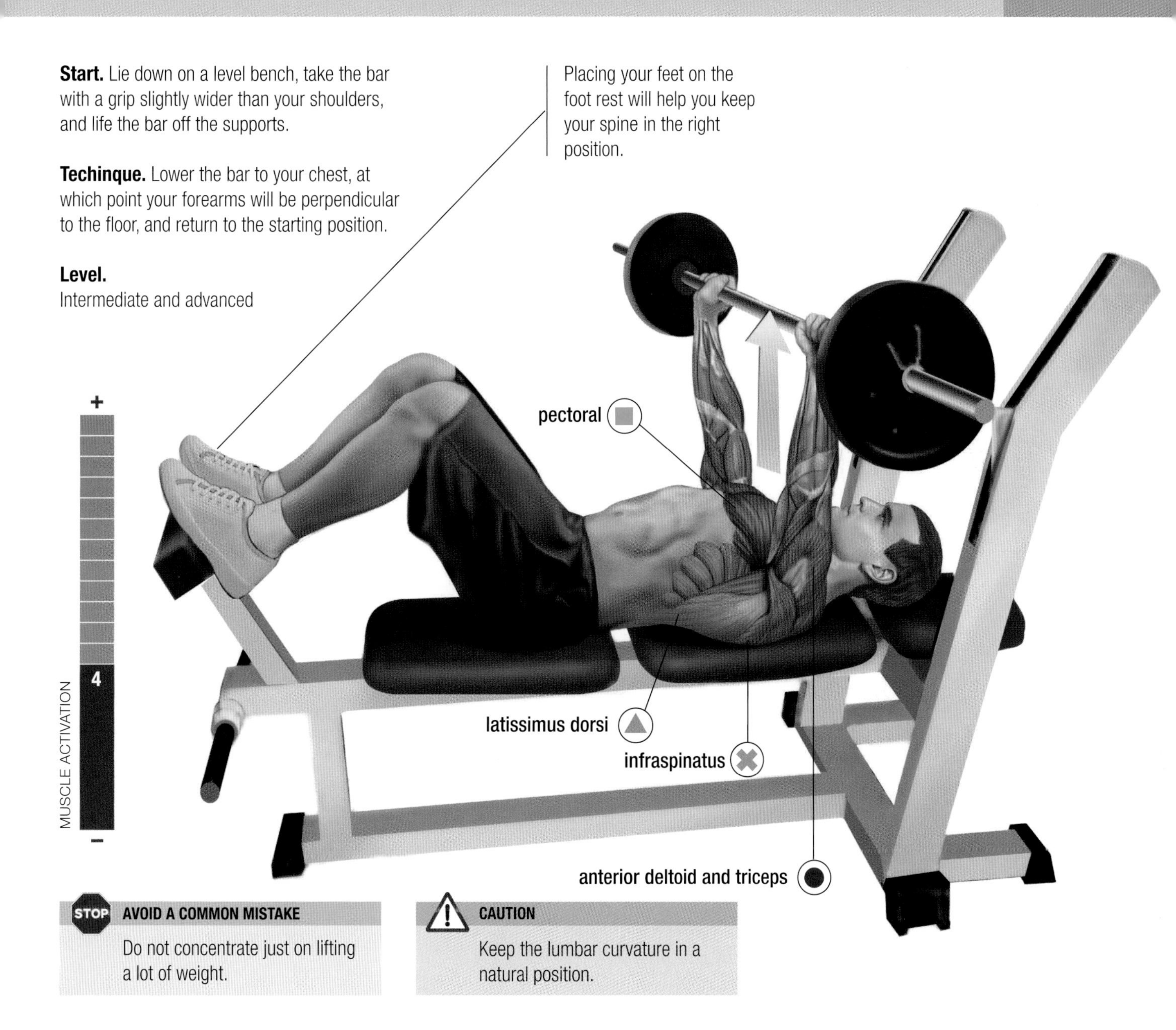
Placing your feet on the foot rest will help you keep your spine in the right position.
MUSCLE ACTIVATION
+
4
–
pectoral
latissimus dorsi
infraspinatus
anterior deltoid and triceps

AVOID A COMMON MISTAKE

Do not concentrate just on lifting a lot of weight.

CAUTION

Keep the lumbar curvature in a natural position.

VARIANT

start
finish

Pectoral

Parallel Bar Dips

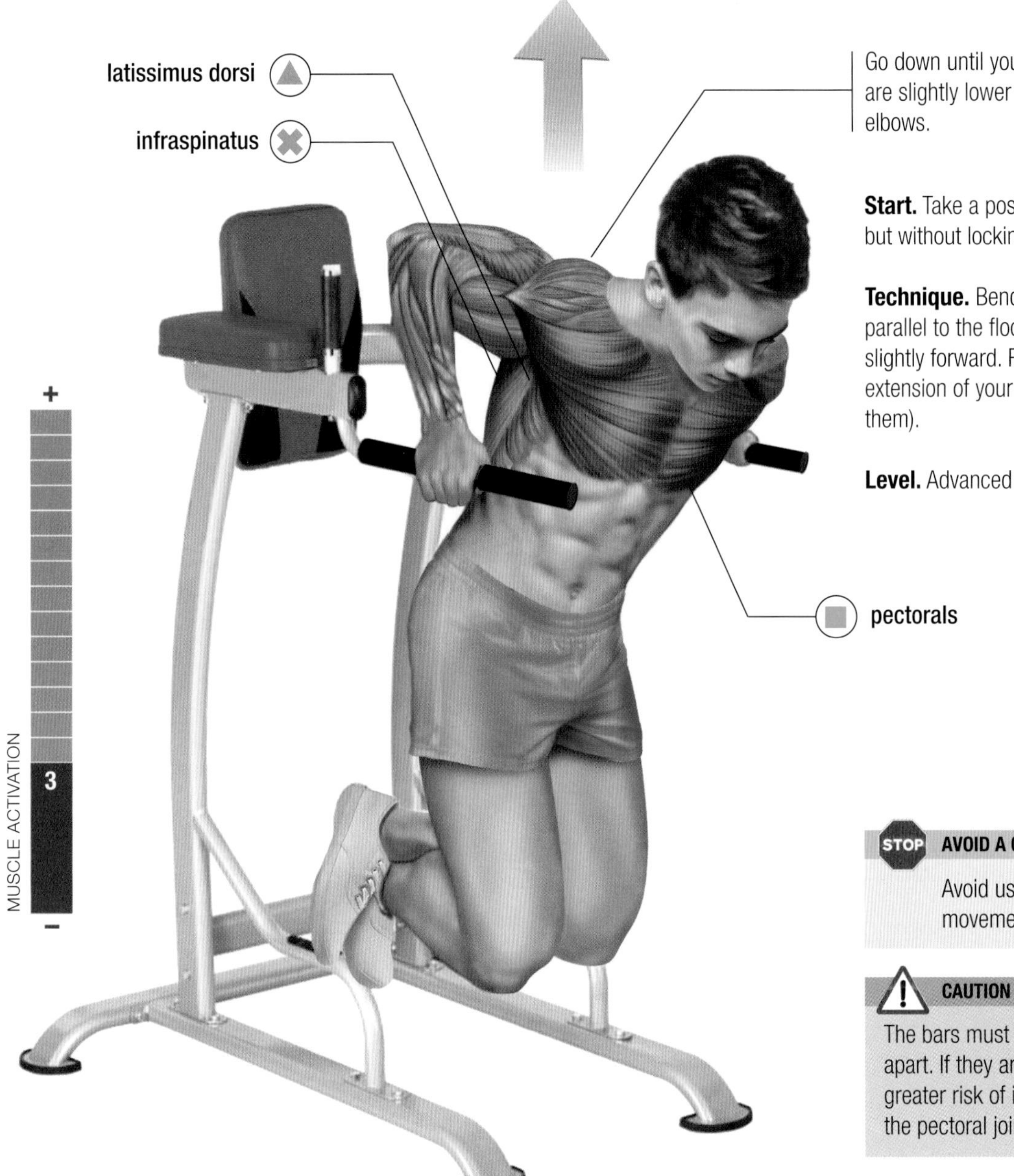

Go down until your shoulders are slightly lower than your elbows.

Start. Take a position on the parallel bars, but without locking the elbows.

Technique. Bend your arms until they are parallel to the floor and your torso leans slightly forward. Rise to nearly complete extension of your elbows (without locking them).

Level. Advanced

AVOID A COMMON MISTAKE

Avoid using very short movements.

CAUTION

The bars must be the right distance apart. If they are too far apart, there is a greater risk of injury to the spot where the pectoral joins the humerus.

VARIANT

Incline Dumbbell Fly

Start. Take a position on an inclined bench, with two dumbbells nearly touching above your chest.

Technique. Keep your elbows slightly bent and lower the dumbbells in line with the medium area of your pectorals. Then reverse the movement.

Level. Advanced

STOP AVOID A COMMON MISTAKE

Don't lower your elbows too far.

CAUTION

Choose the right dumbbell weight for the number of repetitions.

VARIANT

Pectoral

Flat Dumbbell Fly

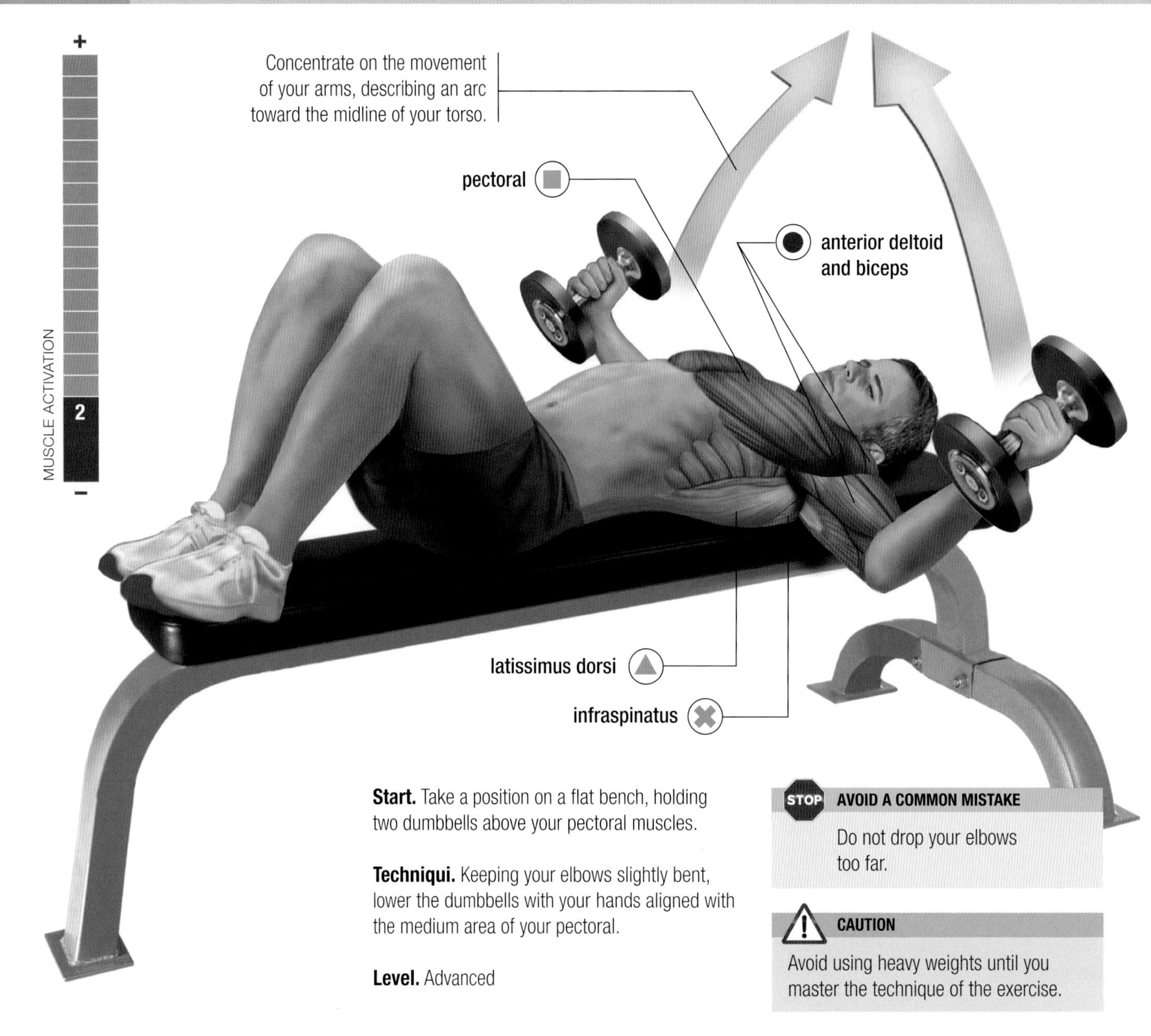
+
MUSCLE ACTIVATION
2
–
Concentrate on the movement of your arms, describing an arc toward the midline of your torso.
pectoral
anterior deltoid and biceps
latissimus dorsi
infraspinatus

Start. Take a position on a flat bench, holding two dumbbells above your pectoral muscles.

Techniqui. Keeping your elbows slightly bent, lower the dumbbells with your hands aligned with the medium area of your pectoral.

Level. Advanced

AVOID A COMMON MISTAKE

Do not drop your elbows too far.

CAUTION

Avoid using heavy weights until you master the technique of the exercise.

VARIANT

start
finish

Incline Bench Press

Start. Get into position on an incline bench and raise the bar over your pectorals with your arms straight and your hands slightly greater than shoulder width apart.

Technique. Lower the bar to the upper part of your chest (where the clavicle and sternum meet) and return to the starting position.

Level. Advanced

AVOID A COMMON MISTAKE
Don't incline the bench too steeply—that is, don't incline it more than 40°.

anterior deltoids and triceps
pectoral
infraspinatus
latissimus dorsi
Lift your chest, keep your shoulders back, and remember to maintain but not accentuate the natural lumbar curvature.

+
3
MUSCLE ACTIVATION
–

CAUTION
Keep your shoulder blades against the bench.

VARIANT

start
finish

Pectoral

Decline Bench Press

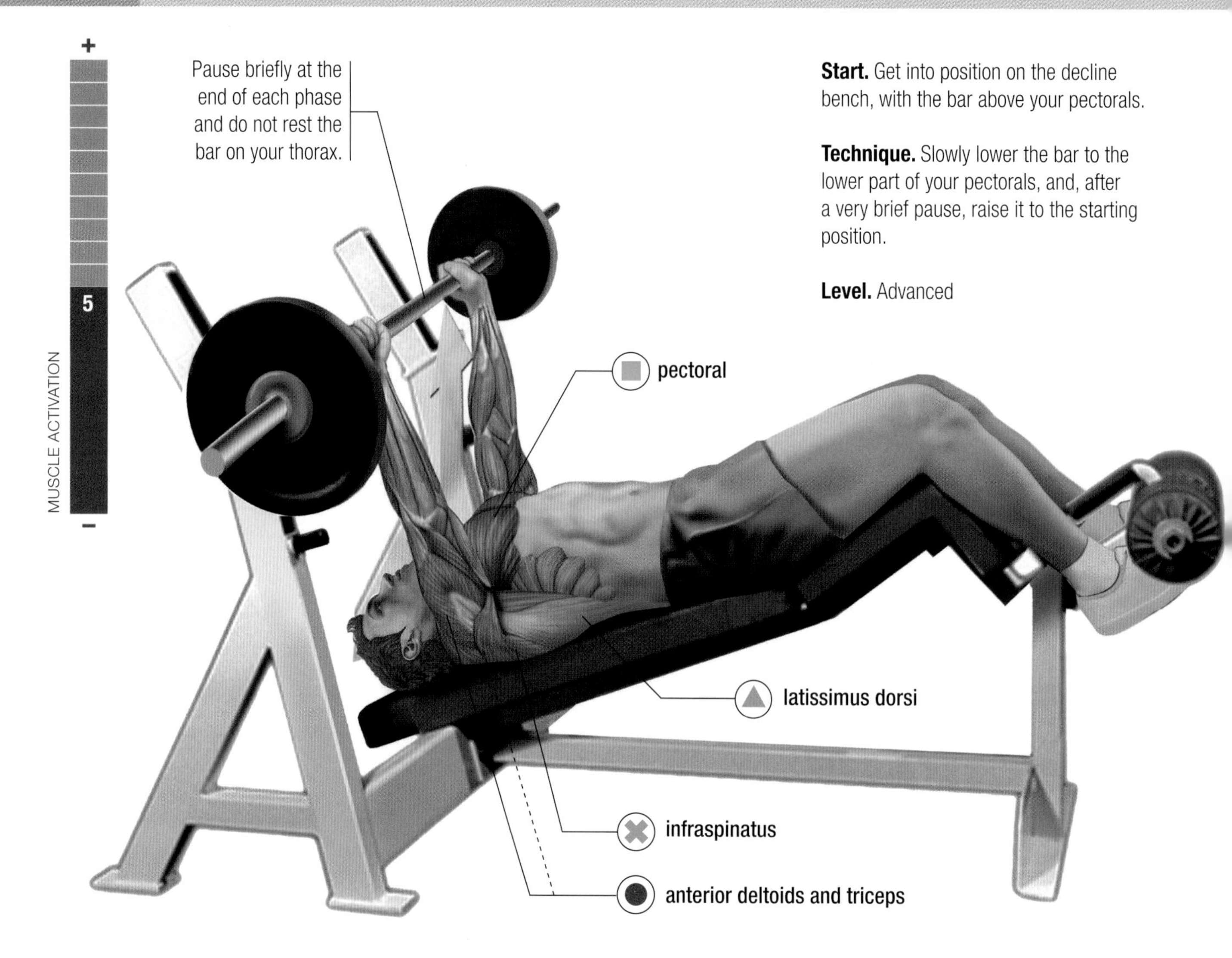
+
5
MUSCLE ACTIVATION
–
Pause briefly at the end of each phase and do not rest the bar on your thorax.
pectoral
latissimus dorsi
infraspinatus
anterior deltoids and triceps

Start. Get into position on the decline bench, with the bar above your pectorals.

Technique. Slowly lower the bar to the lower part of your pectorals, and, after a very brief pause, raise it to the starting position.

Level. Advanced

AVOID A COMMON MISTAKE

Don't lift your gluteals from the bench.

CAUTION

Do not lower the bar onto your neck; this could increase the risk of injury to your shoulder joint.

VARIANT

start
finish

Peck-deck

Start. Get into position on the peck-deck machino and placc your forcarms on the pads.

Technique. Move your arms until the pads touch in front of you. After a brief pause, slowly perform the reverse movement and return to the starting position.

Level. Beginner, intermediate, and advanced

STOP AVOID A COMMON MISTAKE

Avoid separating your elbows from the pads during movement.

CAUTION

Do not raise your hips or exert too much pressure with your hands.

Pectoral

Dumbbell Pull-over

AVOID A COMMON MISTAKE

Avoid extending your elbow joint instead of extending your shoulder joint.

CAUTION

Do not arch your back too much during the descent and the initial part of raising the dumbbell. Take particular care with the lumbar region.

Start. Lie down on a bench while holding a dumbbell with both hands above your chest. Maintain a slight bend in your elbows.

Technique. Slowly lower the dumbbell behind your head, making an arc, until your arms are parallel to the floor and in line with your torso. Then perform the reverse movement by raising the dumbbell to the starting position, with your arms perpendicular to your torso.

Level. Advanced

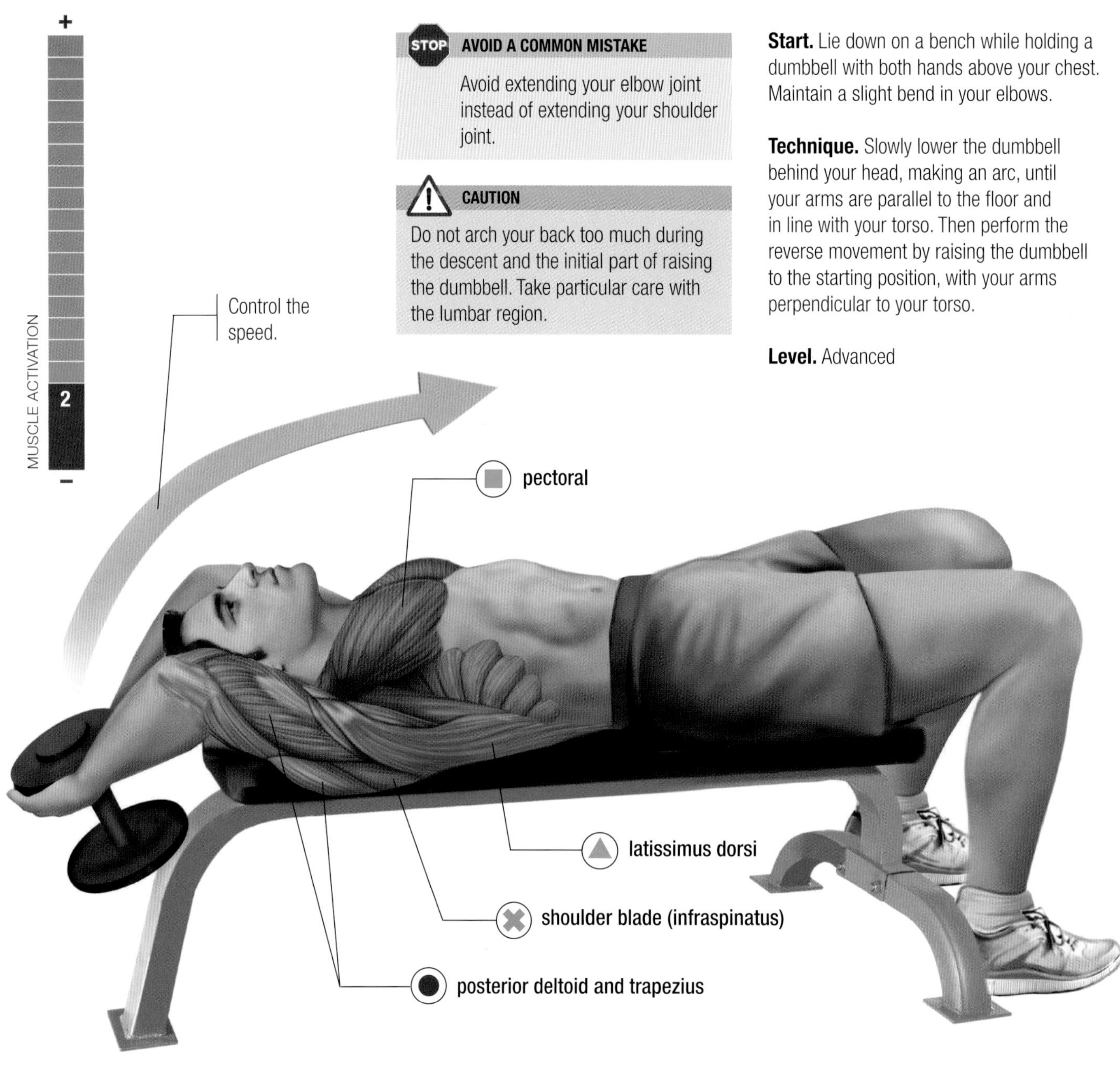

VARIANT

Bench Press on the Machine

Start. Sit on the pectoral press machine, with your shoulder blades pressed against the back cushion and your arms parallel to the floor.

Technique. Raise both arms together, slowly and evenly, toward the front by straightening your elbows. Your elbows remain just beneath shoulder level. When you reach total extension of your elbows, avoid locking the joint. After a brief pause, slowly return to the starting point.

Level. Beginner, intermediate, and advanced

AVOID A COMMON MISTAKE

Keep your shoulder blades pressed against the back cushion.

CAUTION

Avoid pressing against the back cushion with your head and exerting pressure with your neck.

Back

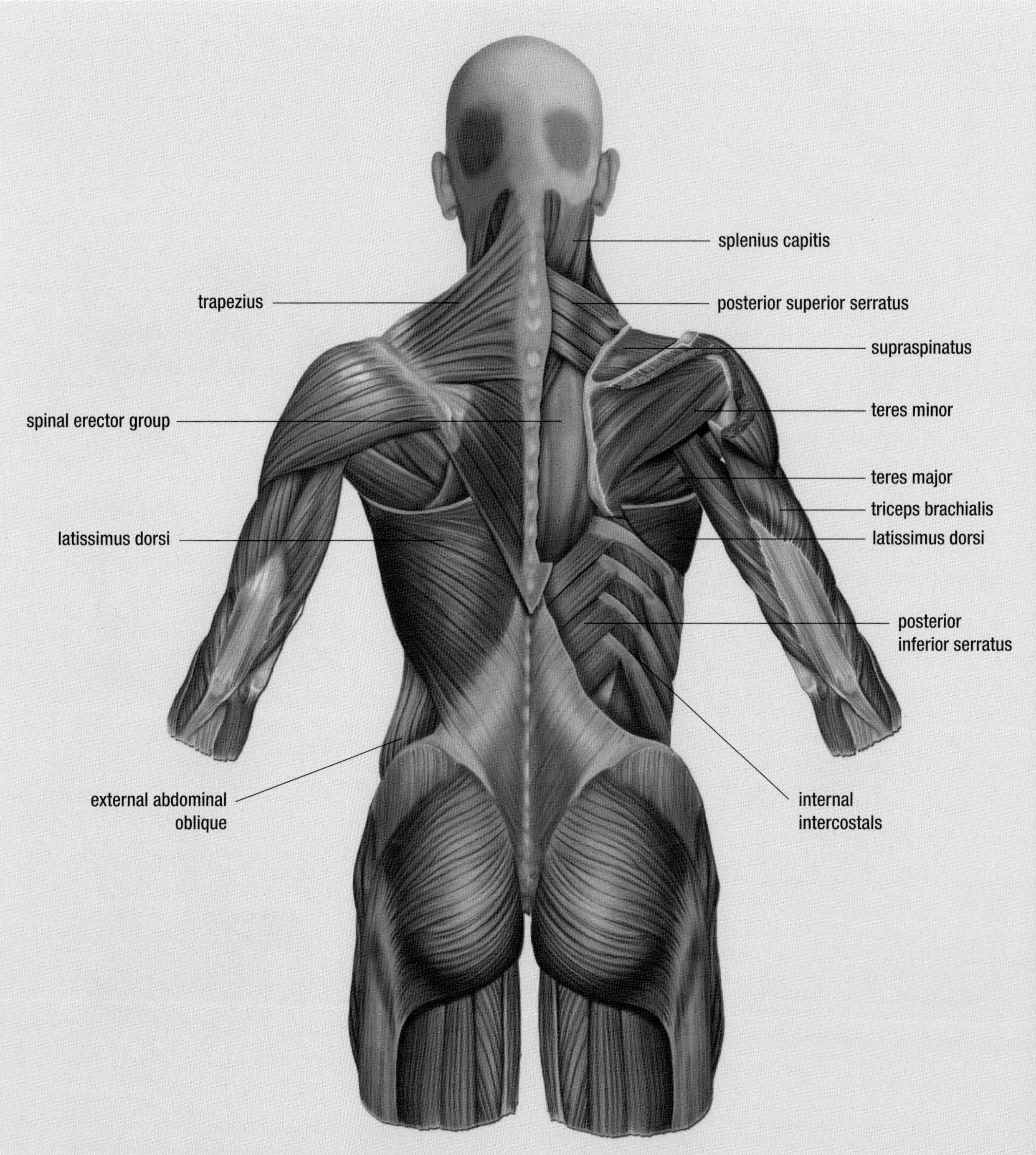
splenius capitis
trapezius
posterior superior serratus
supraspinatus
spinal erector group
teres minor
teres major
triceps brachialis
latissimus dorsi
latissimus dorsi
posterior
inferior serratus
external abdominal
oblique
internal
intercostals

The back includes the entire posterior section of the torso. It also contains two powerful muscles, namely the latissimus dorsi and the trapezius, and smaller ones, such as the quadratus lumborum, the teres major, the teres minor, the rhomboid, the infraspinatus, the supraspinatus, the spinal erector, and others. Many bodybuilders pay less attention to the back muscles, because they are less conspicuous than the pectoral and the bicep muscles. This is a mistake, because these muscles are essential for balancing the body and maintaining good posture.

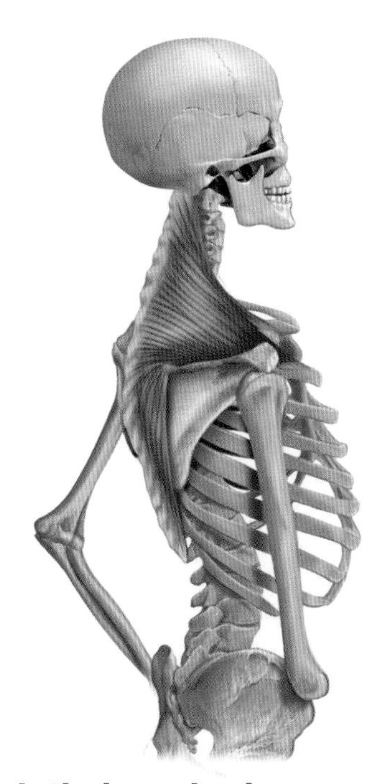

Trapezius

This muscle arises in the occipital bone and the spiny apophyses of the cervical and thoracic vertebrae, and it inserts in the acromion and the spine of the scapula. The trapezius has various functions, because its upper fibers produce the elevation in the scapula; its middle fibers, the adduction of the scapula; and the inner ones, the depression and adduction of the scapula. This muscle thus contributes to sports such as archery, rowing disciplines, and Greco–Roman wrestling.

semispinalis
splenius capitis
serratus posterior superior
transversospinalis
external intercostals
longissimus thoracis
serratus posterior inferior
rotators
abdominal internal oblique
quadratus lumborum

Latissimus dorsi

This muscle arises in the spiny apophyses of T6 to L5 and the sacrum, and in the posterior ridge of the ilium. Its insertion is in the third proximal of the humerus. It gives breadth to the back below the shoulders and its function is to extend the shoulders, so it participates in large measure in sports such as climbing, rowing, swimming, and judo. It is justly referred to as one of the climbing muscles.

Quadratus Lumborum

This muscle arises in the ilium (crest and inner edge), and it inserts in the lower edge of the 12th rib and in the transverse apophyses of vertebrae L1 to L4. Its main functions are the elevation of the pelvis and bending the torso to the side, so it is used especially in the sports of rowing and rhythmic gymnastics.

Trapezius

Dumbbell Shrugs

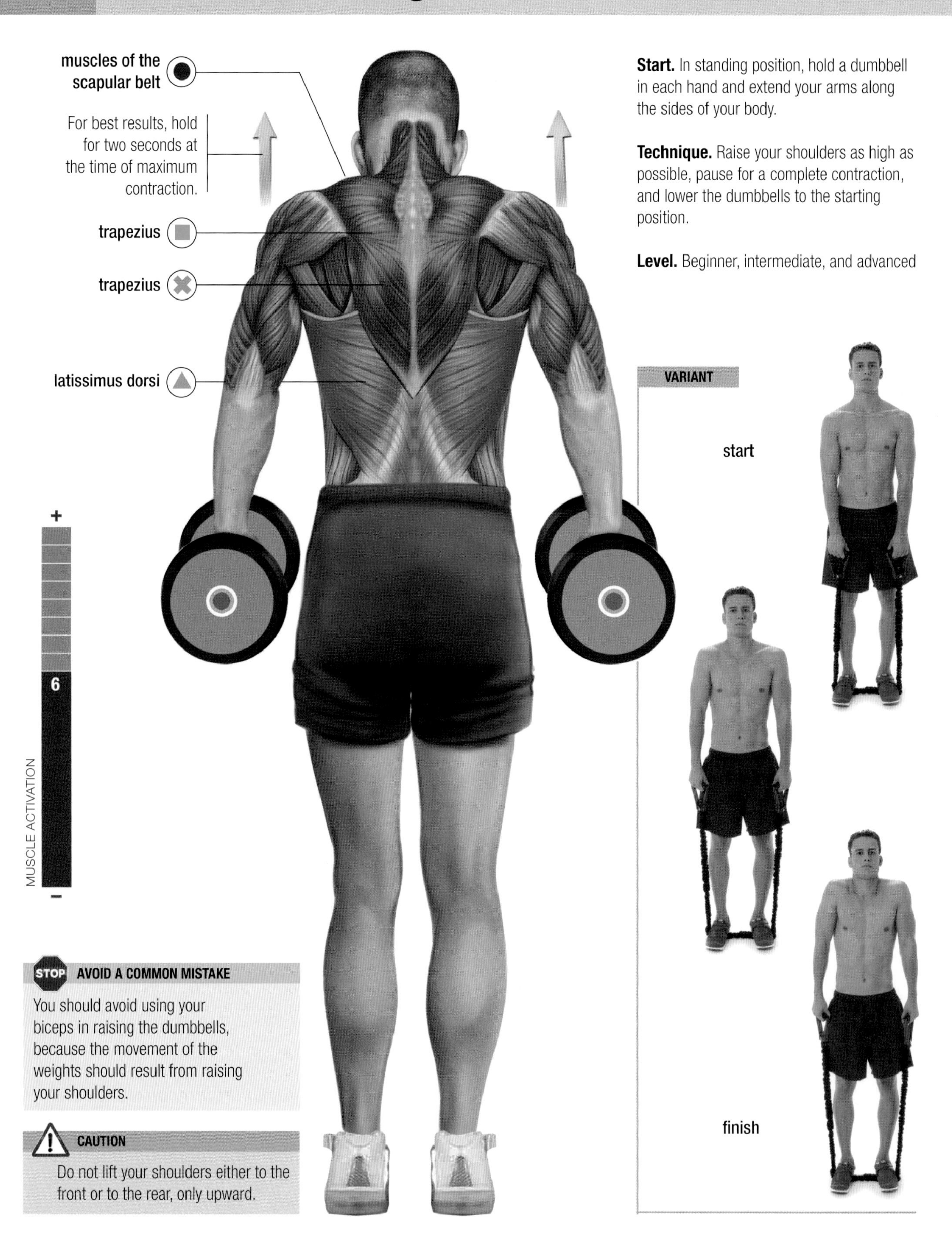

Start. In standing position, hold a dumbbell in each hand and extend your arms along the sides of your body.

Technique. Raise your shoulders as high as possible, pause for a complete contraction, and lower the dumbbells to the starting position.

Level. Beginner, intermediate, and advanced

AVOID A COMMON MISTAKE

You should avoid using your biceps in raising the dumbbells, because the movement of the weights should result from raising your shoulders.

CAUTION

Do not lift your shoulders either to the front or to the rear, only upward.

Upright Rowing

Start. Hold the bar with your hands palms-down.

Technique. Raise the bar toward your chin with your elbows held high and then lower it to the starting point, all in a smooth motion.

Level. Advanced

Keeping your knees slightly bent keeps your lumbar region from getting involved, thus avoiding back problems.

AVOID A COMMON MISTAKE

The first few times an athlete performs this exercise, it is common to lower the elbows below the hands. Avoid this mistake: always keep your elbows above your hands.

CAUTION

Using lots of weight in upright rowing can lead to shoulder problems. Limit the weights, especially if you are a newcomer to lifting.

Latissimus dorsi

Pull-ups

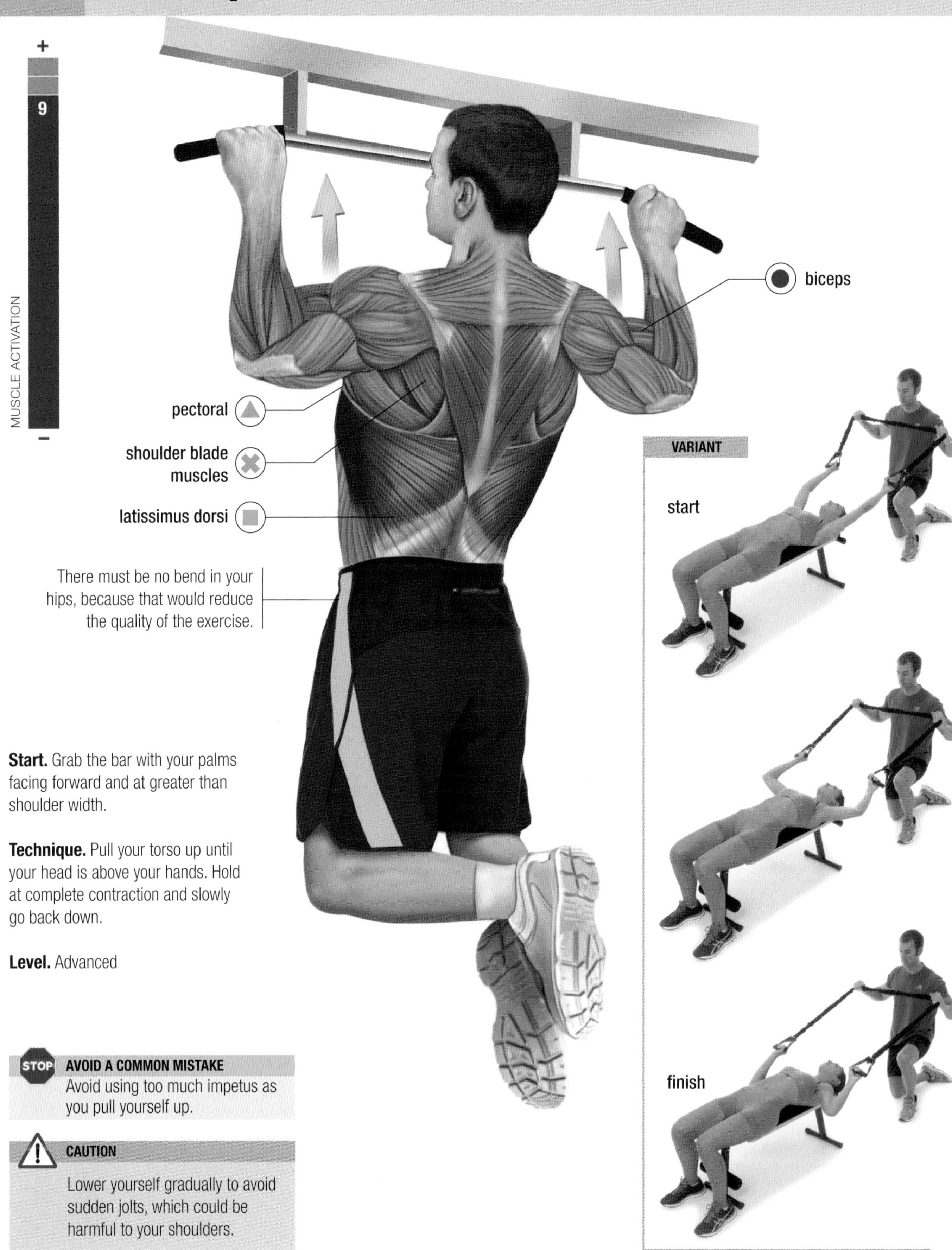

Start. Grab the bar with your palms facing forward and at greater than shoulder width.

Technique. Pull your torso up until your head is above your hands. Hold at complete contraction and slowly go back down.

Level. Advanced

AVOID A COMMON MISTAKE
Avoid using too much impetus as you pull yourself up.

CAUTION

Lower yourself gradually to avoid sudden jolts, which could be harmful to your shoulders.

Reverse-grip Pull-down

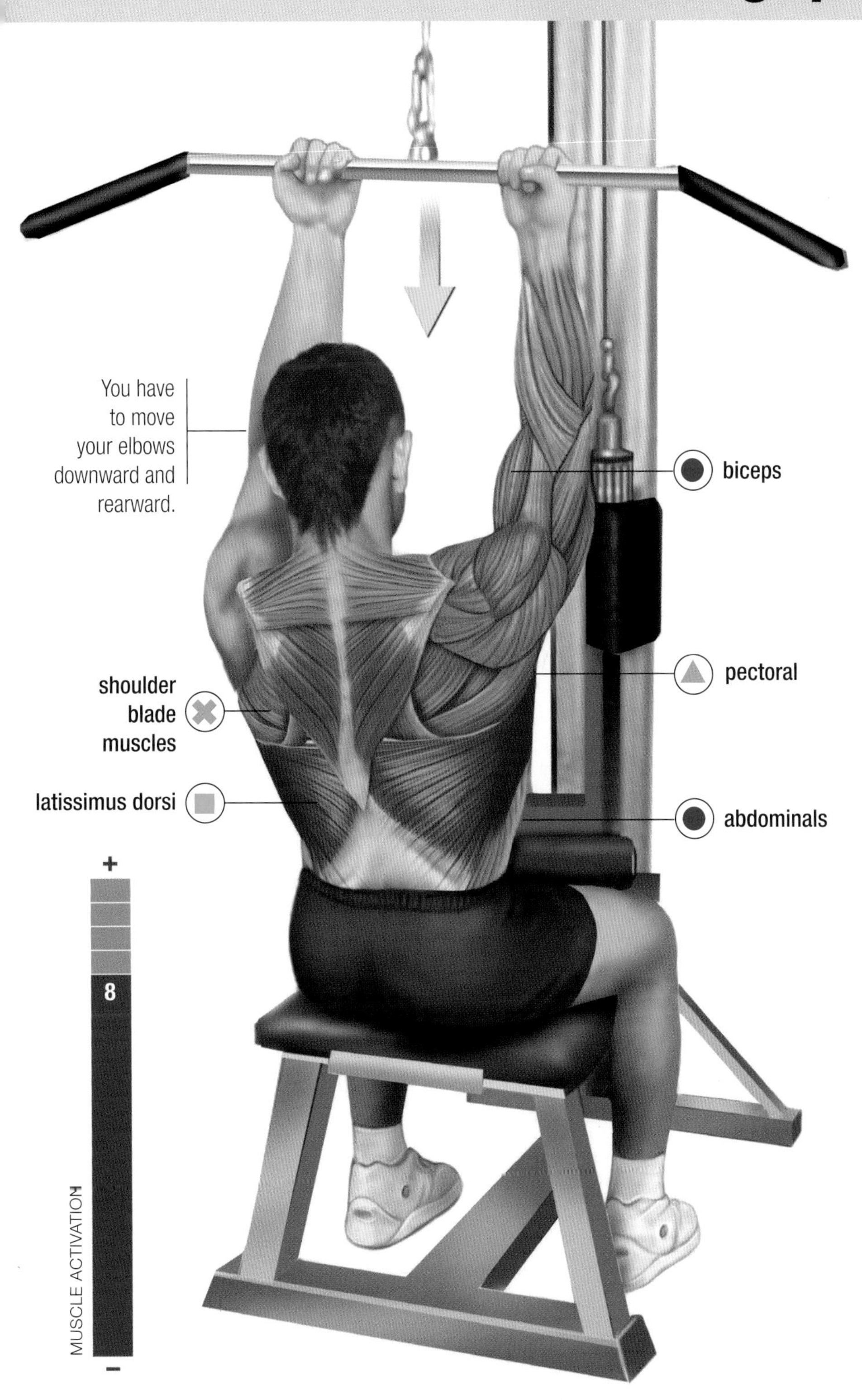

Start. Sit on the machine and hold the bar with your palms facing you and about 12 in. (30 cm) apart. Anchor your knees beneath the support cushions.

Technique. Pull the bar down toward your chest as far as the top part of your pectorals, inclining your torso no more than 30º. Hold the bar down and return to the starting position.

Level. Intermediate and advanced

VARIANT

AVOID A COMMON MISTAKE

Raising your shoulders and sinking your chest is one common mistake, so take special care to avoid this deterioration of technique.

CAUTION

Not anchoring your knees may produce pain in the lumbar area and limit your capacity for work.

Latissimus dorsi

Front Pull-down

Start. Sit on the machine and hold the bar with your palms facing forward, slightly wider than shoulder width apart. Don't forget to anchor your knees on the pull-down machine.

Technique. Pull the bar down toward your chest and move your shoulders downward and rearward. Hold the bar down and then return to the starting position.

VARIANT

Level. Intermediate and advanced

AVOID A COMMON MISTAKE

Anchor your legs to avoid raising the thorax.

CAUTION

Don't bend or extend your neck.

Horizontal Pull

Start. Hold the cable of the lower pulley and sit on a bench facing it. Adjust the seat in such a way that, when you put your feet on the support, your knees are slightly bent and your torso is vertical.

Technique. Pull your hands toward your torso, hold the contraction, and slowly return to the starting position.

Level. Intermediate and advanced

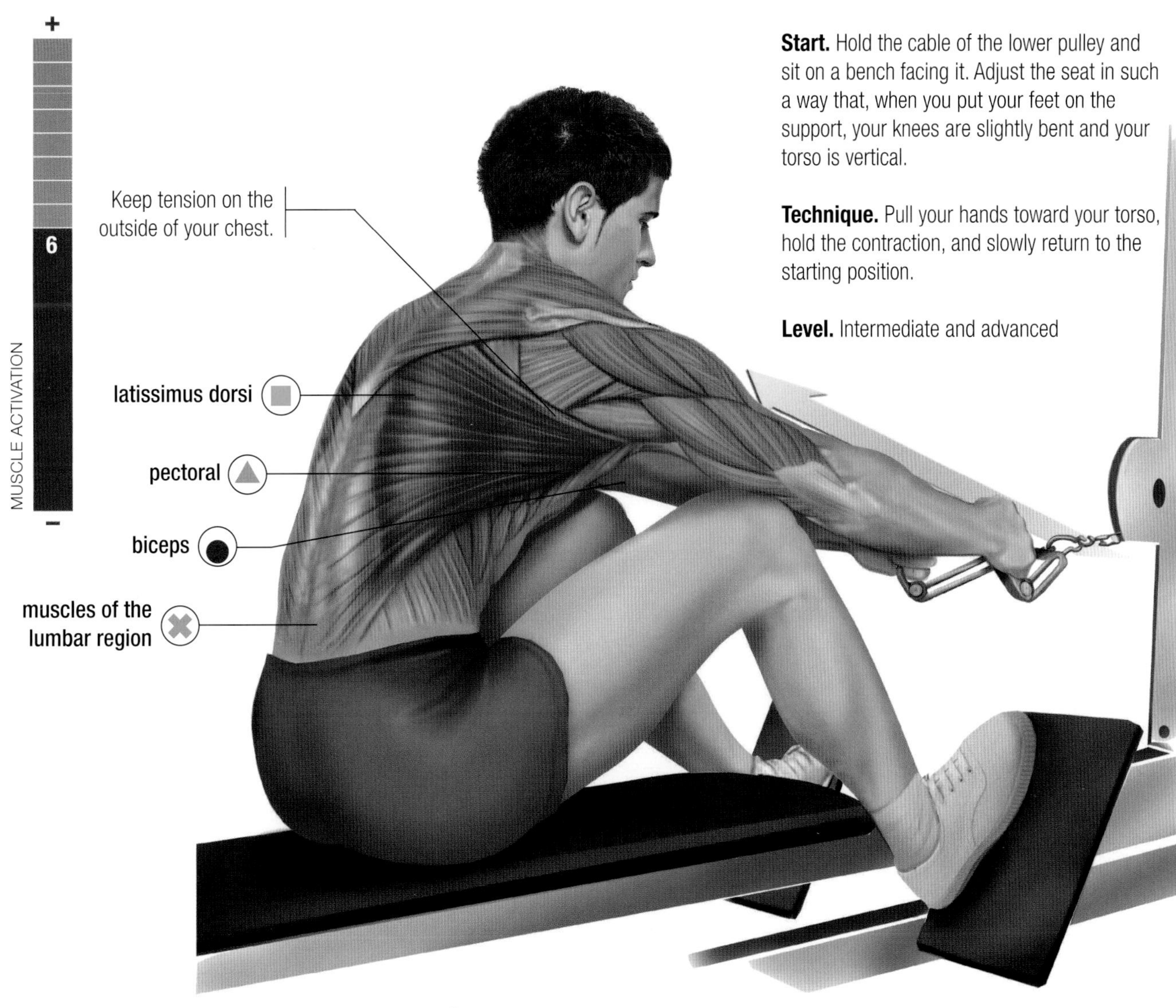

AVOID A COMMON MISTAKE

Avoid raising your shoulders.

CAUTION

Maintain the physiological curvature of your spine.

VARIANT

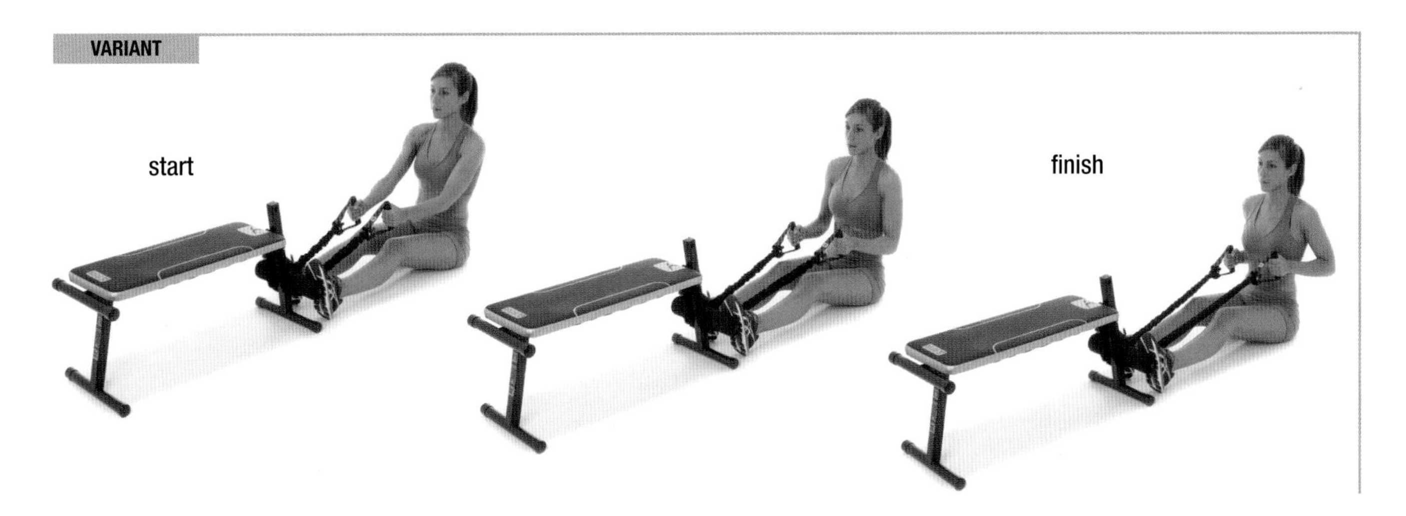

Latissimus dorsi

Rowing on the Machine

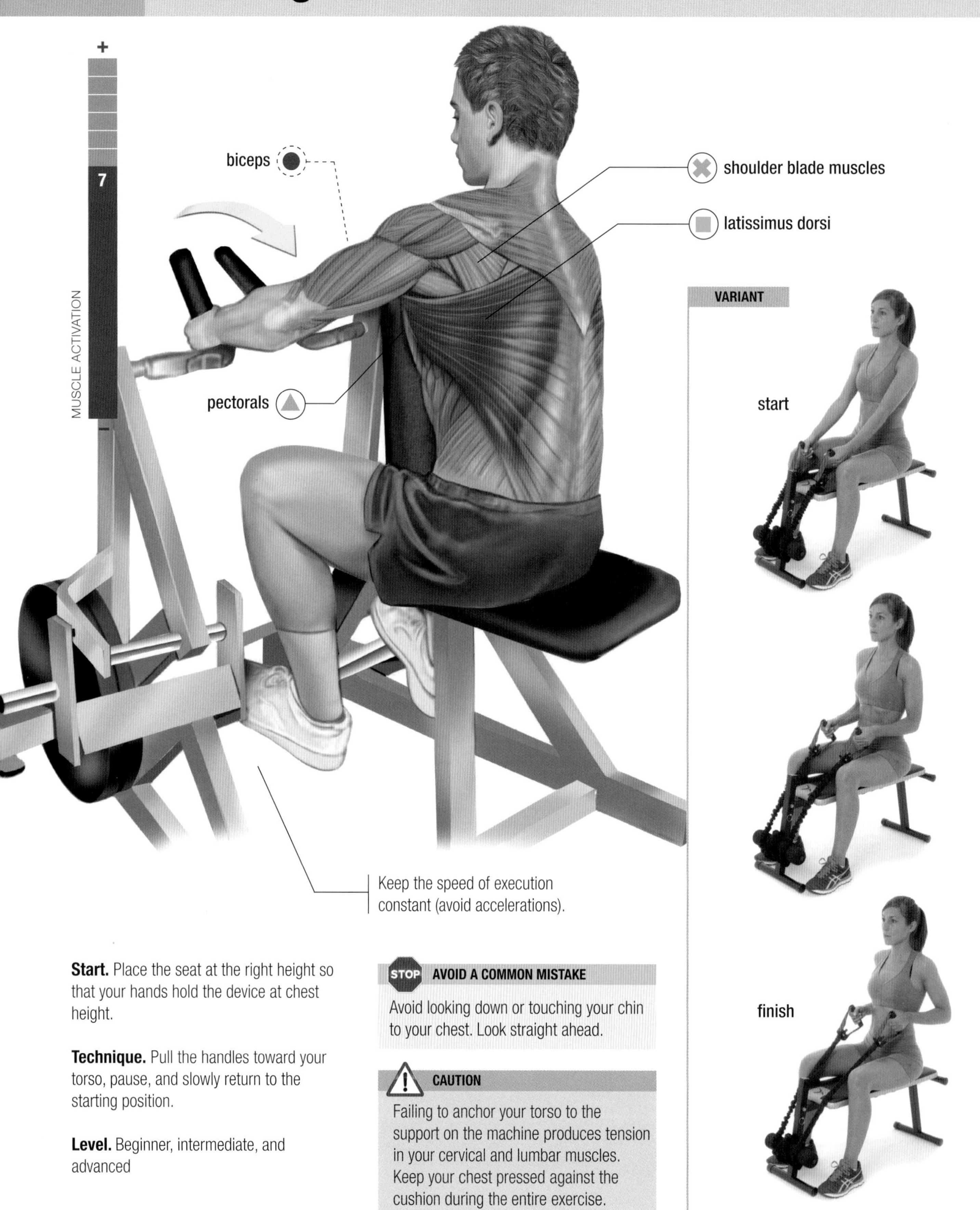

Start. Place the seat at the right height so that your hands hold the device at chest height.

Technique. Pull the handles toward your torso, pause, and slowly return to the starting position.

Level. Beginner, intermediate, and advanced

STOP AVOID A COMMON MISTAKE

Avoid looking down or touching your chin to your chest. Look straight ahead.

CAUTION

Failing to anchor your torso to the support on the machine produces tension in your cervical and lumbar muscles. Keep your chest pressed against the cushion during the entire exercise.

Pull-overs on the Machine

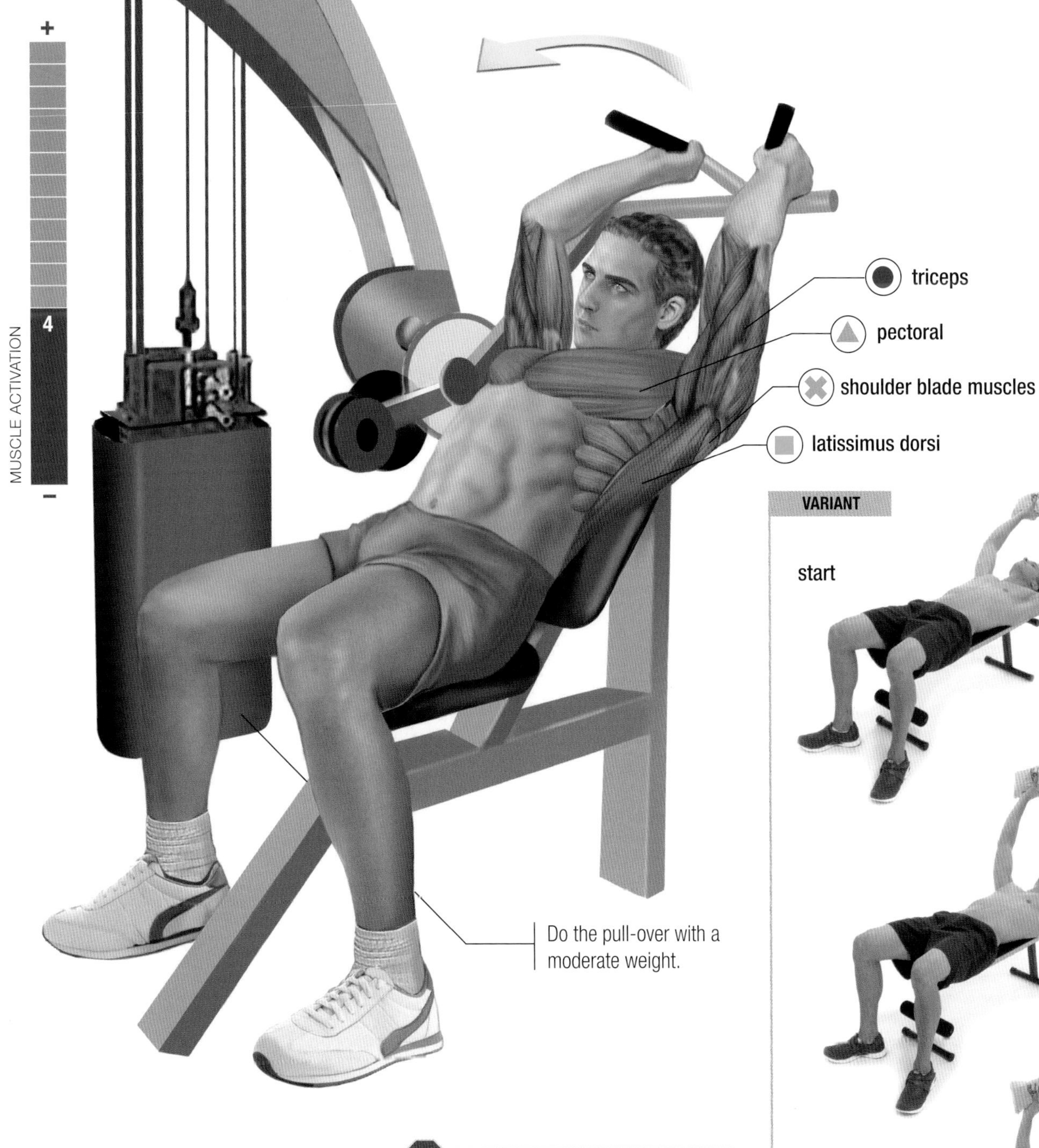

VARIANT

Start. Sit on the pull-over machine and grab the handles with your arms bent.

Technique. Push the handles downward (in front of your chest), hold them down, and raise them back to the starting position.

Level. Beginner, intermediate, and advanced

AVOID A COMMON MISTAKE

One common error is changing the natural lumbar curvature. Avoid doing this, because it may lead to back pain and problems.

CAUTION

Remember that this is an exercise for your latissimus dorsi, so avoid direct involvement of your trapezius.

Latissimus dorsi

Straight-arm Pull-downs

Start. Take a standing position with your feet slightly more than shoulder width apart and your knees slightly bent.

Technique. Move your arms from high to low, forming an arc that goes from about 95º toward the legs. Pause and return to the starting position.

Level. Advanced

AVOID A COMMON MISTAKE

Avoid bending your knees.

CAUTION

Do not straighten your legs: this would risk producing excess tension in your lumbar region.

Pull-downs with a V-grip

AVOID A COMMON MISTAKE

It is common to sink your chest and exert pressure with your abdominals to move a heavy weight. Avoid this mistake and reduce the weight if necessary to produce a good execution.

CAUTION

Maintain the physiological lumbar curvature and avoid using too much weight.

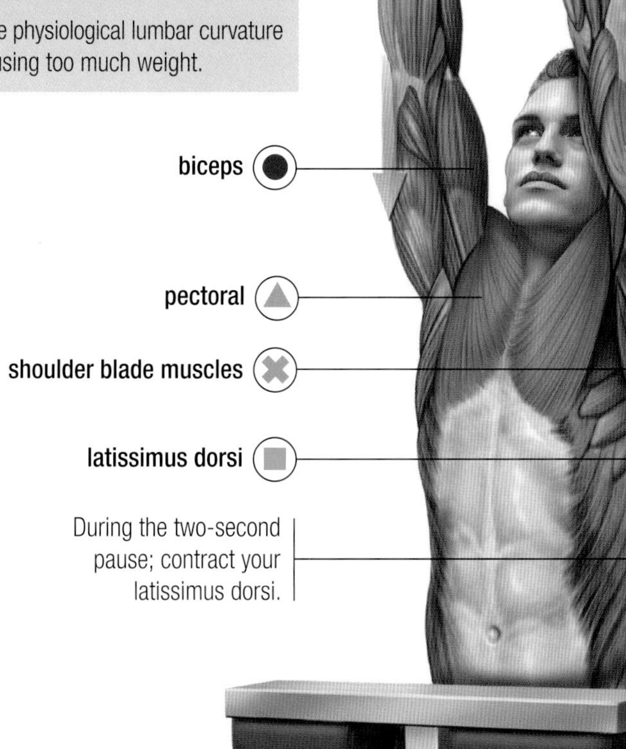

Start. Sit down facing the machine with your knees anchored against the support cushions and pull the handle to your sternum, slightly leaning your upper body to the rear.

Technique. Draw the V-handle toward your chest, hold, and return to the starting position.

Level. Intermediate and advanced

VARIANT

Latissimus dorsi

Dumbbell Row

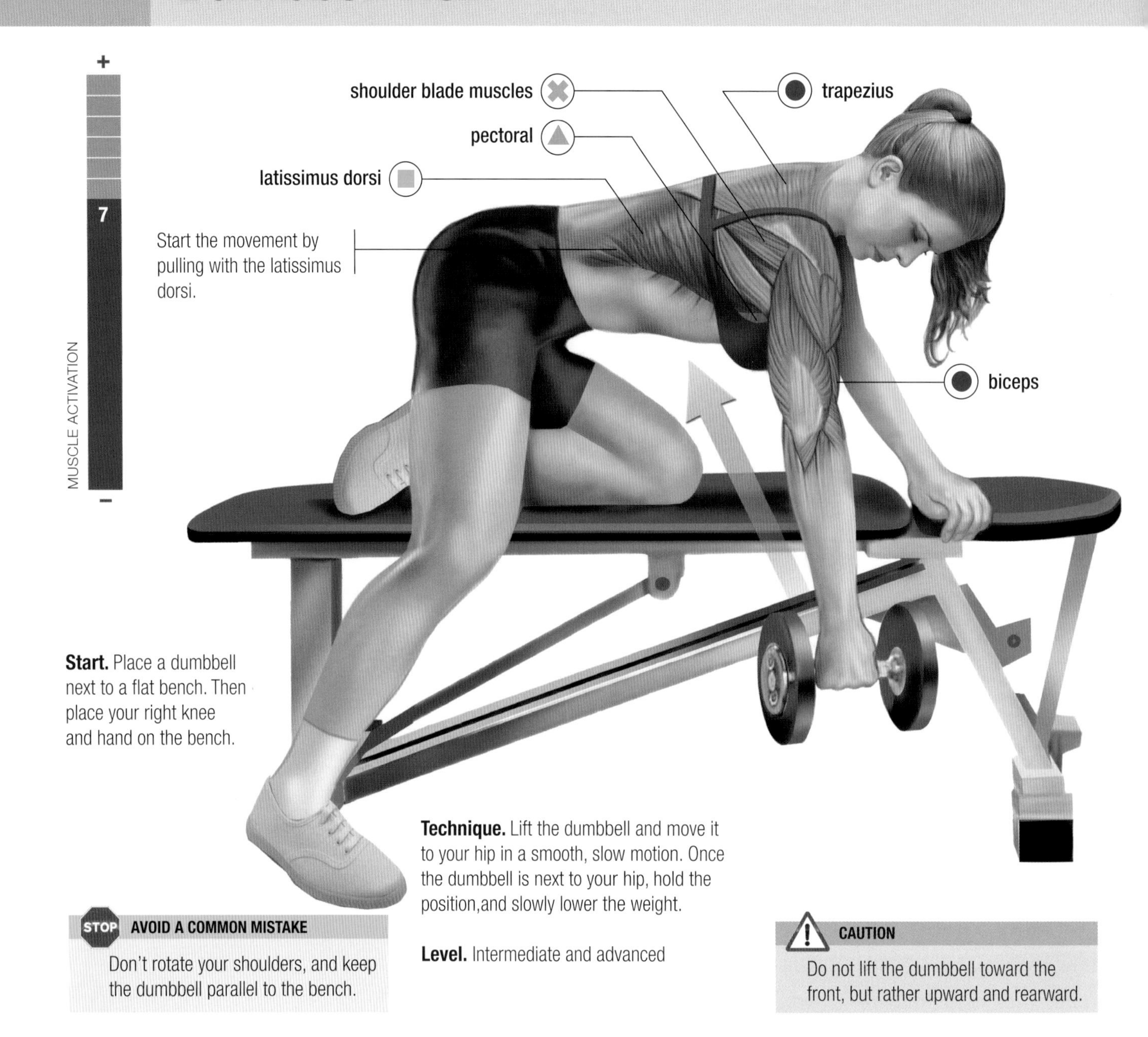

Start. Place a dumbbell next to a flat bench. Then place your right knee and hand on the bench.

Technique. Lift the dumbbell and move it to your hip in a smooth, slow motion. Once the dumbbell is next to your hip, hold the position,and slowly lower the weight.

Level. Intermediate and advanced

AVOID A COMMON MISTAKE

Don't rotate your shoulders, and keep the dumbbell parallel to the bench.

CAUTION

Do not lift the dumbbell toward the front, but rather upward and rearward.

VARIANT

Pull-overs on the Machine

AVOID A COMMON MISTAKE

Do not let the weight fall back quickly in the descent phase, because the sudden jolt could damage your shoulders.

CAUTION

Avoid arching your back during the exercise.

Start. Place a flat bench in front of a pulley. Lie down with your head toward the pulley and grip the long bar. Stretch your arms toward the rear and in line with your torso.

Technique. Pull upward and forward by describing an arc that ends when your hands are over the lower part of your chest. Then return to the starting position.

Level. Advanced

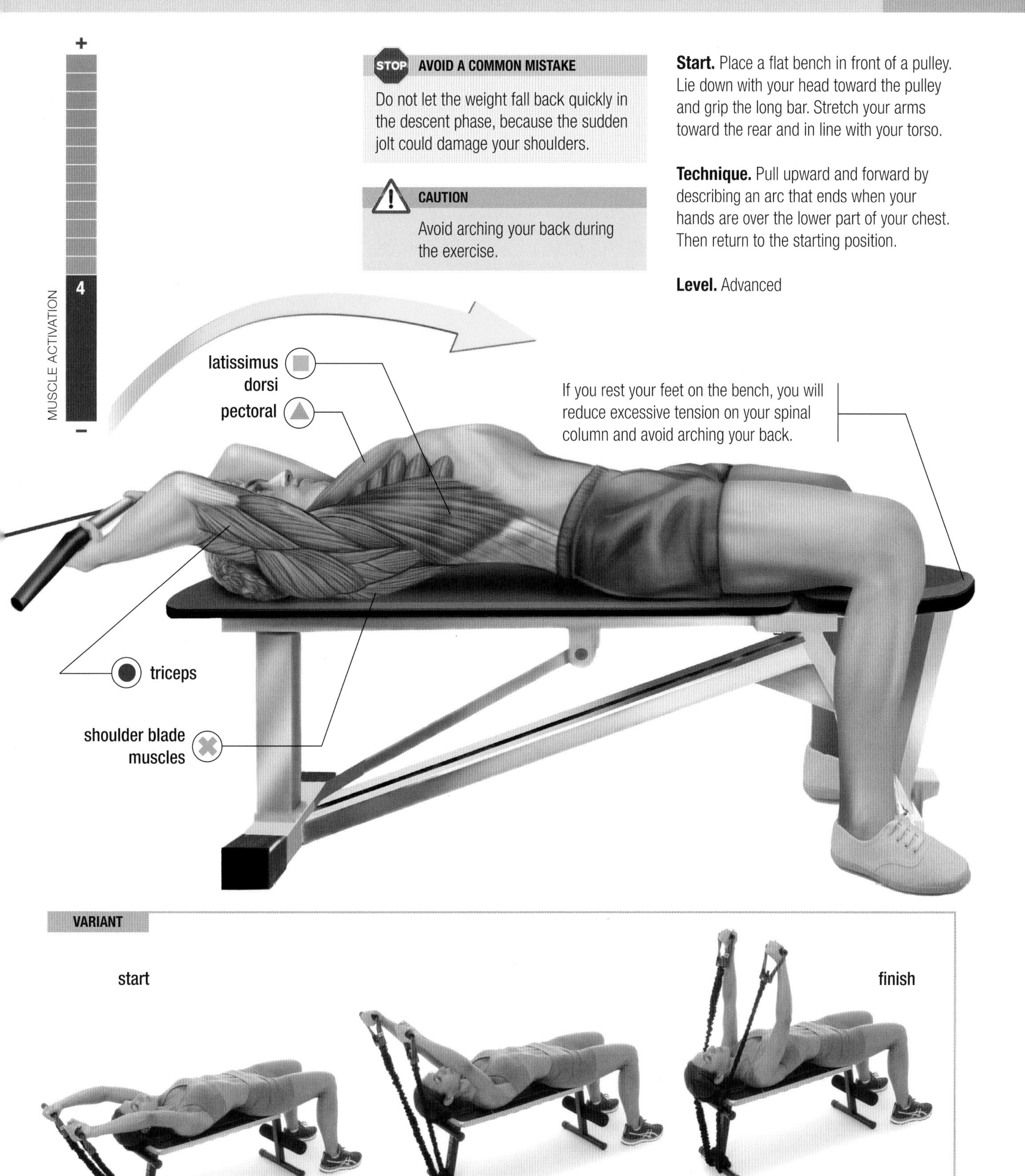
STOP
MUSCLE ACTIVATION
+
4
–
latissimus dorsi
pectoral
triceps
shoulder blade muscles
If you rest your feet on the bench, you will reduce excessive tension on your spinal column and avoid arching your back.
VARIANT
start
finish

Latissimus dorsi

Inclined Row

Start. Take a stance with your torso slightly inclined toward the front, your back straight, and your knees bent about 90–95º. Next, grab the bar with your hands placed below the plates. You can use a V-grip or an inclined rowing device. In both cases, the grip will be more comfortable and secure.

Technique. Pull the bar upward and hold it near your torso. Then lower it in a controlled manner.

Level. Advanced

AVOID A COMMON MISTAKE

Avoid short movements and work with less weight and broad movements.

CAUTION

Avoid jerking motions, which could increase the risk of back injury. If you are working with a free-weight bar, be sure to anchor the opposite end.

Latissimus dorsi

Barbell Row

Start. Lean your chest forward at an angle of less than 95º with the hips, while holding the bar with the hands shoulder-width apart and the palms facing downward.

Technique. Raise the bar to the lower part of your chest. Pause and then slowly lower the bar until your arms are straight.

Level. Advanced

AVOID A COMMON MISTAKE

Do not jerk or perform quick movements, because this could damage the joints involved.

CAUTION

Keep your back straight and your knees slightly bent.

VARIANT

start

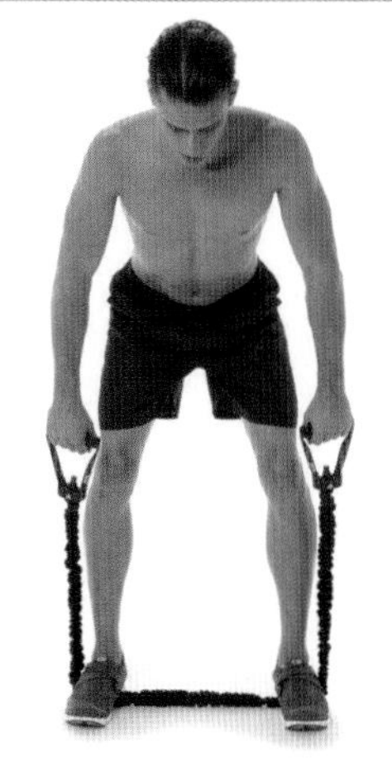

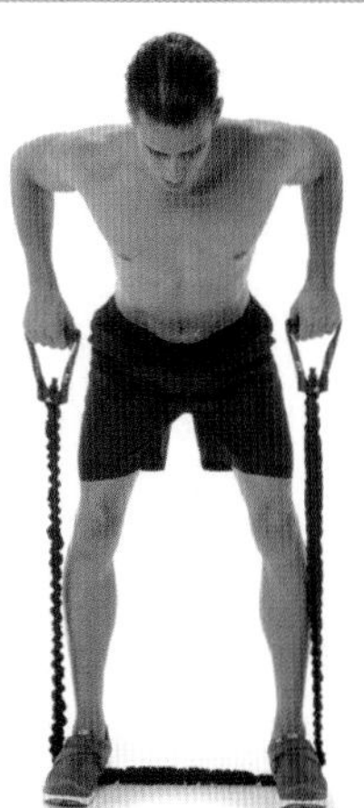

finish

Quadratus lumborum on the Machine

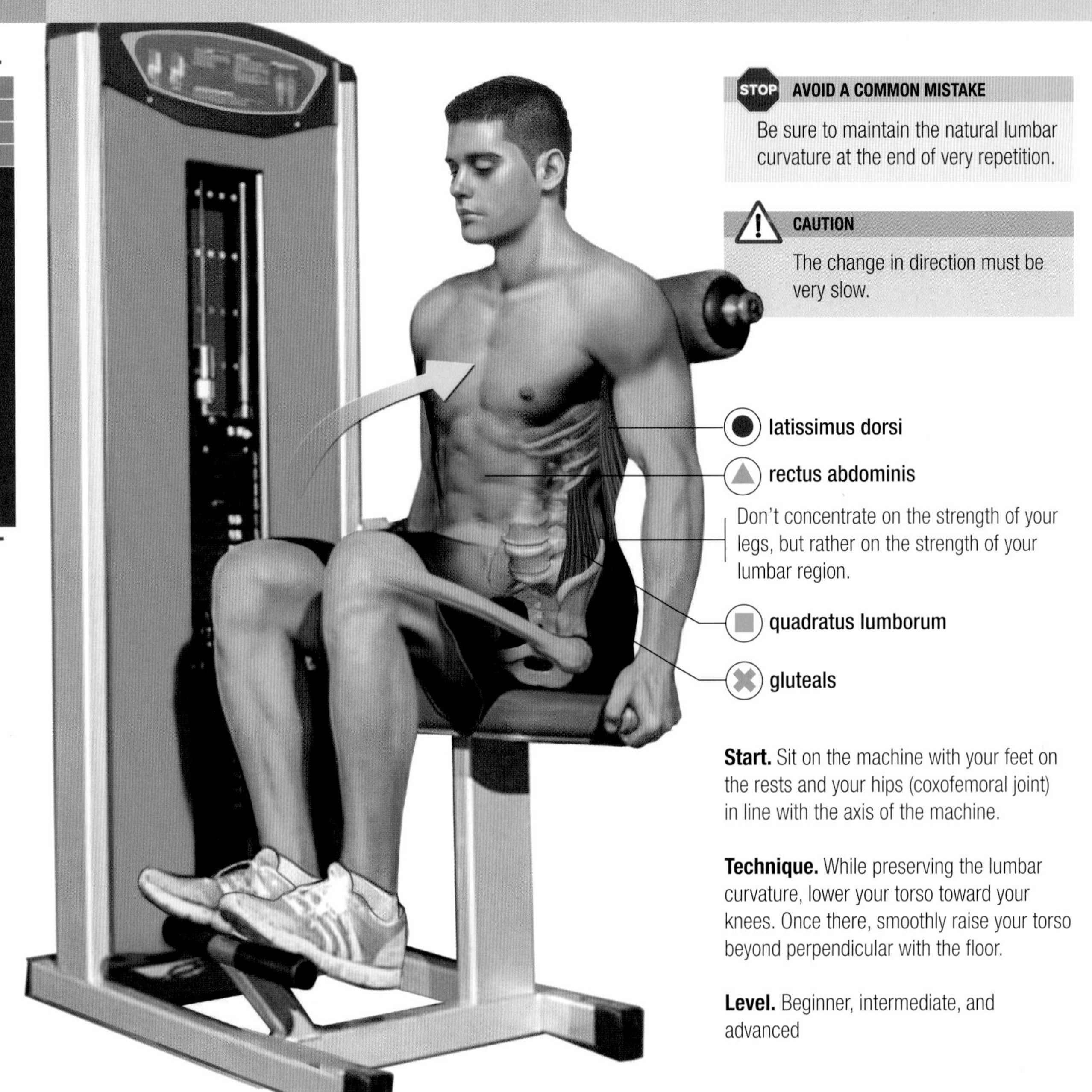

AVOID A COMMON MISTAKE

Be sure to maintain the natural lumbar curvature at the end of very repetition.

CAUTION

The change in direction must be very slow.

Start. Sit on the machine with your feet on the rests and your hips (coxofemoral joint) in line with the axis of the machine.

Technique. While preserving the lumbar curvature, lower your torso toward your knees. Once there, smoothly raise your torso beyond perpendicular with the floor.

Level. Beginner, intermediate, and advanced

VARIANT

Quadratus lumborum on the Bench

Start. Take a position on the machine for the lumbar region, with your feet in the rests and with the front of your thighs leaning against the cushions. Your arms should be crossed on your chest, and you may hold a barbell plate or some other type of weight if you are at the intermediate or advanced level.

Technique. Lower your torso by bending at your hips and moving a relatively short distance. Then raise your torso until it is aligned with your legs.

Level. Intermediate and advanced

STOP AVOID A COMMON MISTAKE

Avoid bending your hips to an angle of less than 70° between your torso and the floor.

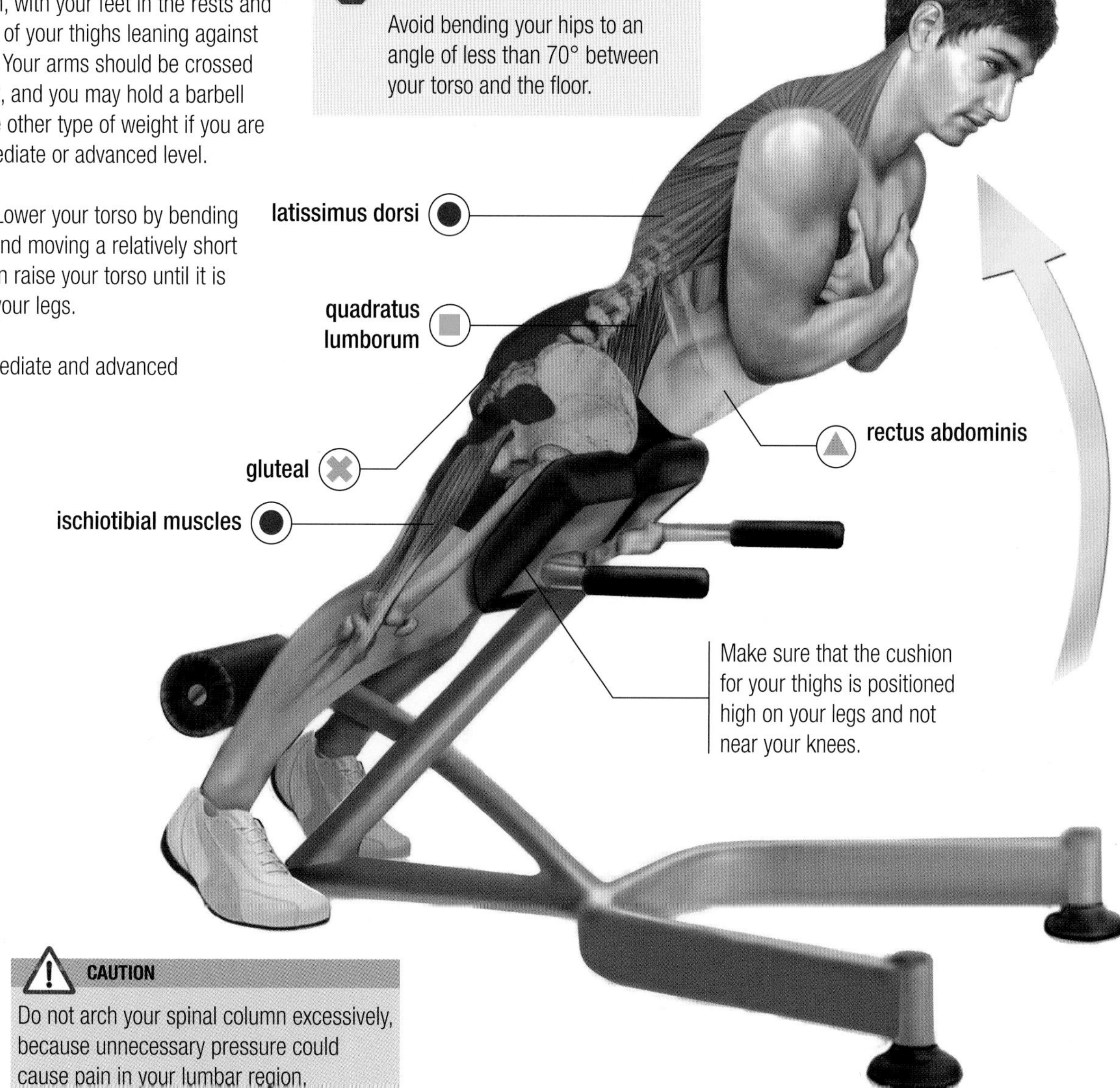

Make sure that the cushion for your thighs is positioned high on your legs and not near your knees.

CAUTION

Do not arch your spinal column excessively, because unnecessary pressure could cause pain in your lumbar region.

VARIANT

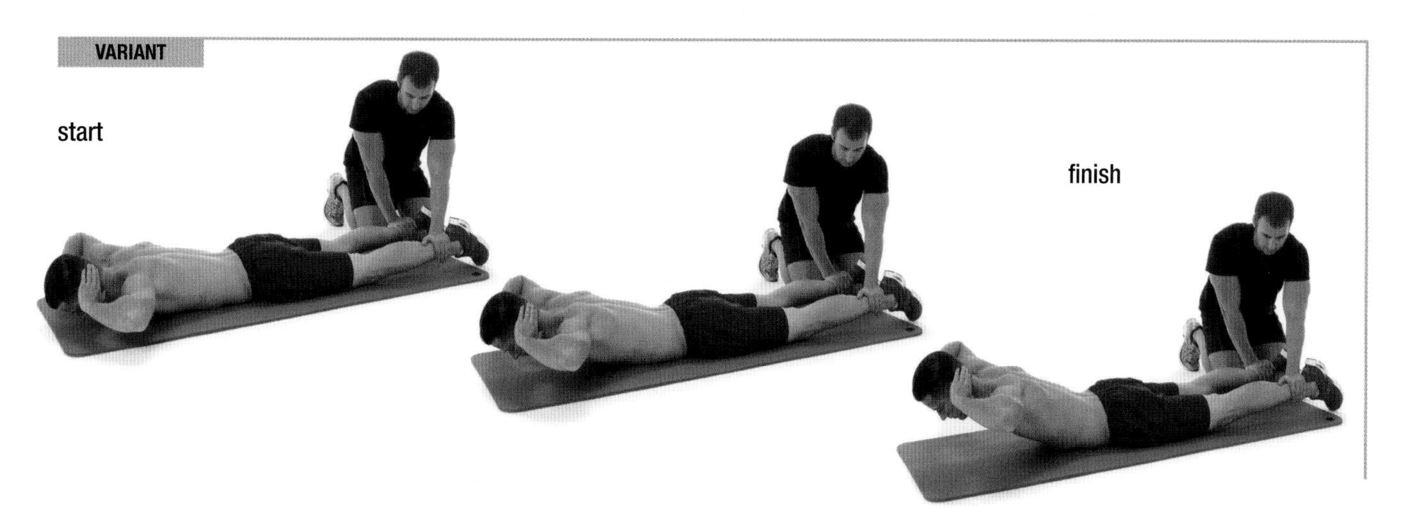

Shoulder

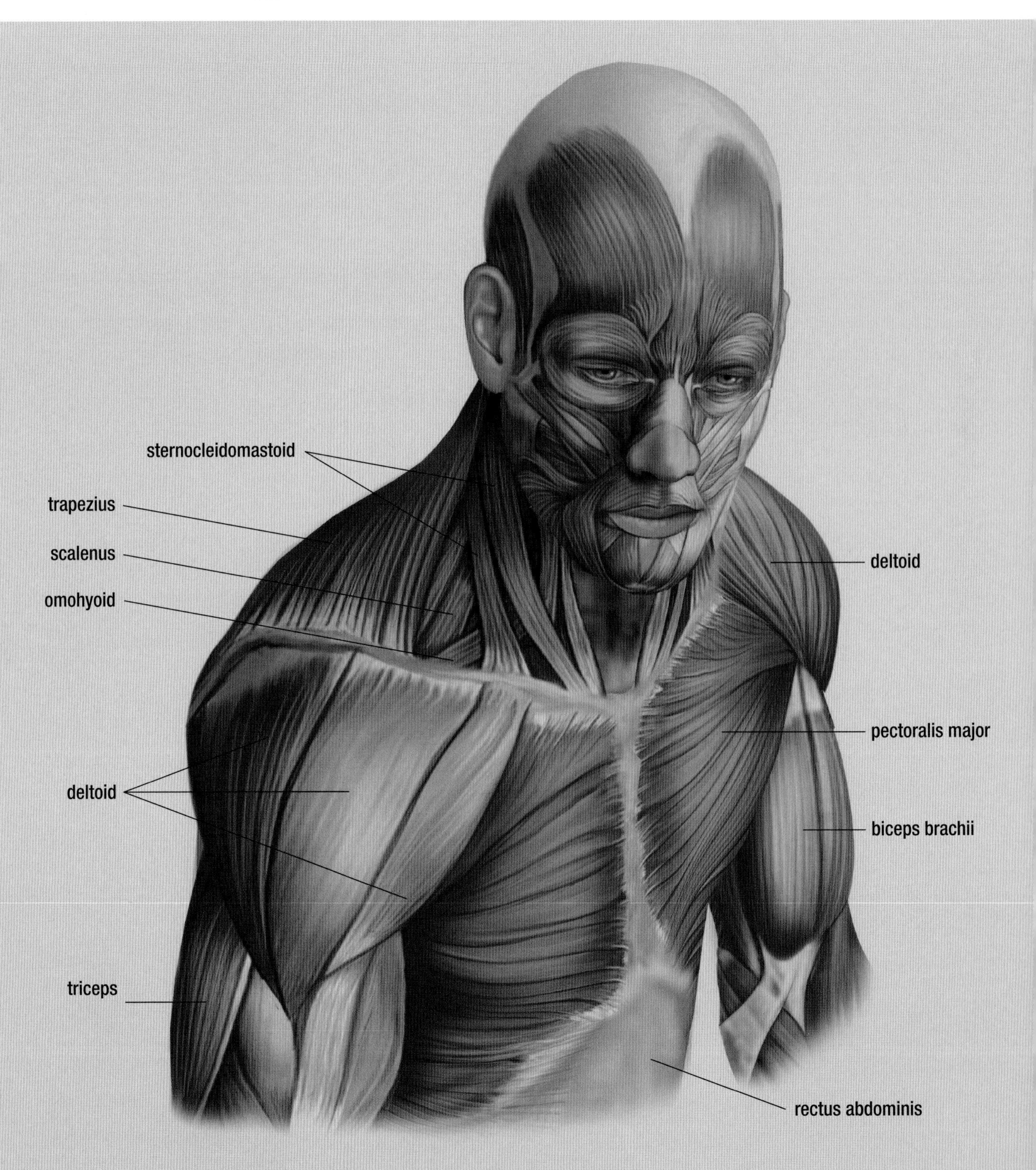

sternocleidomastoid
trapezius
scalenus
omohyoid
deltoid
triceps
deltoid
pectoralis major
biceps brachii
rectus abdominis

The shoulder is the upper lateral portion of the trunk, where the arms meet the torso. The ends of the humerus, the scapula, and the clavicle come together in the shoulder. The scapular–humeral joint permits great mobility; but, at the same time, it is relatively unstable in working with heavy weights or extreme angles, so you must be very careful in all exercises that involve the shoulder. The deltoid is the muscle that covers this joint and gives an appearance of roundness to this part of the body. Its main functions are abduction and antepulsion of the shoulder.

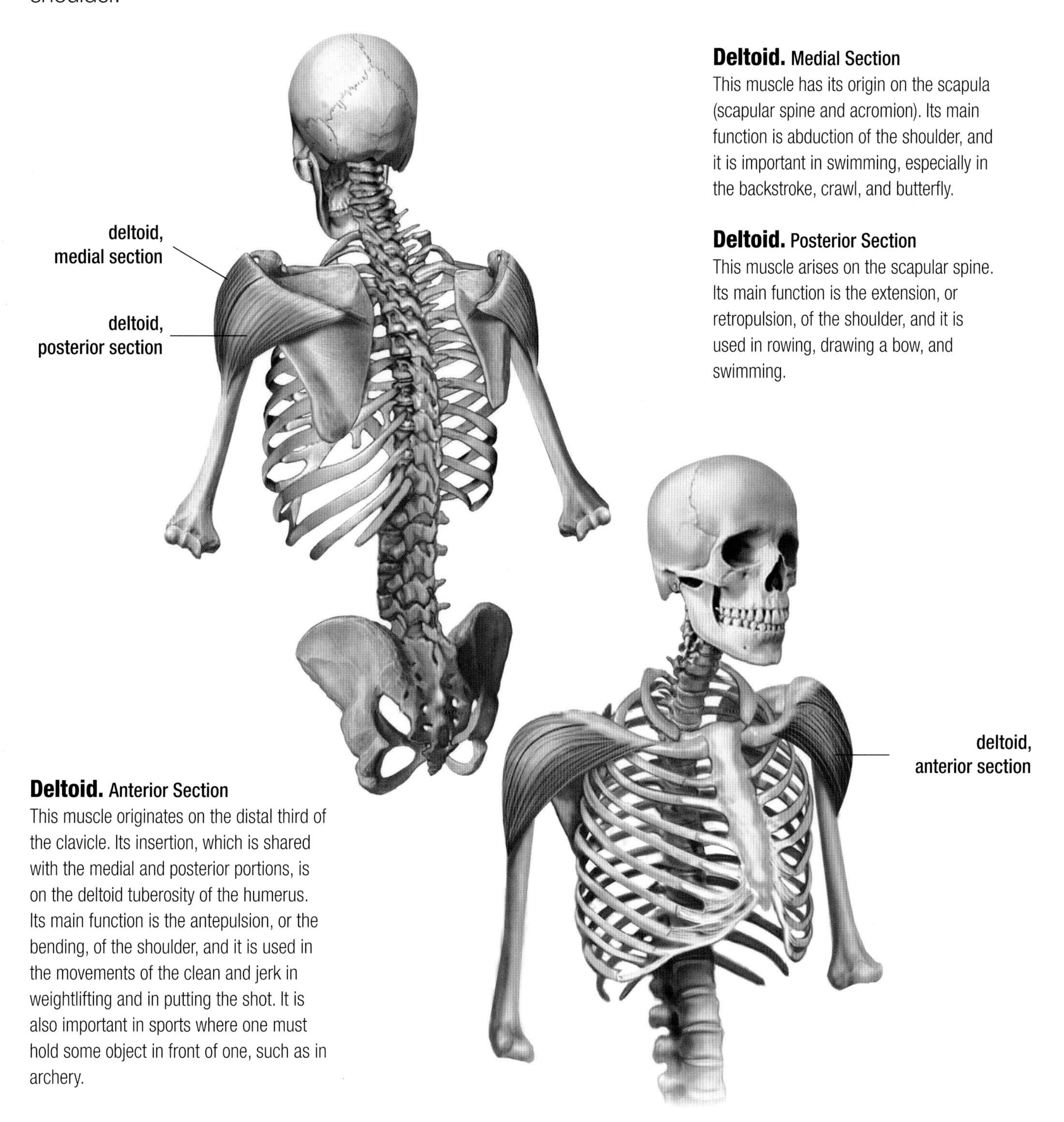

Deltoid. Medial Section

This muscle has its origin on the scapula (scapular spine and acromion). Its main function is abduction of the shoulder, and it is important in swimming, especially in the backstroke, crawl, and butterfly.

Deltoid. Posterior Section

This muscle arises on the scapular spine. Its main function is the extension, or retropulsion, of the shoulder, and it is used in rowing, drawing a bow, and swimming.

Deltoid. Anterior Section

This muscle originates on the distal third of the clavicle. Its insertion, which is shared with the medial and posterior portions, is on the deltoid tuberosity of the humerus. Its main function is the antepulsion, or the bending, of the shoulder, and it is used in the movements of the clean and jerk in weightlifting and in putting the shot. It is also important in sports where one must hold some object in front of one, such as in archery.

Deltoid

Lateral Raises

CAUTION

Avoid leaning your torso while lowering the dumbbells, and avoid jerking with your back as you lift them.

AVOID A COMMON MISTAKE

Avoid bending the elbows too much.

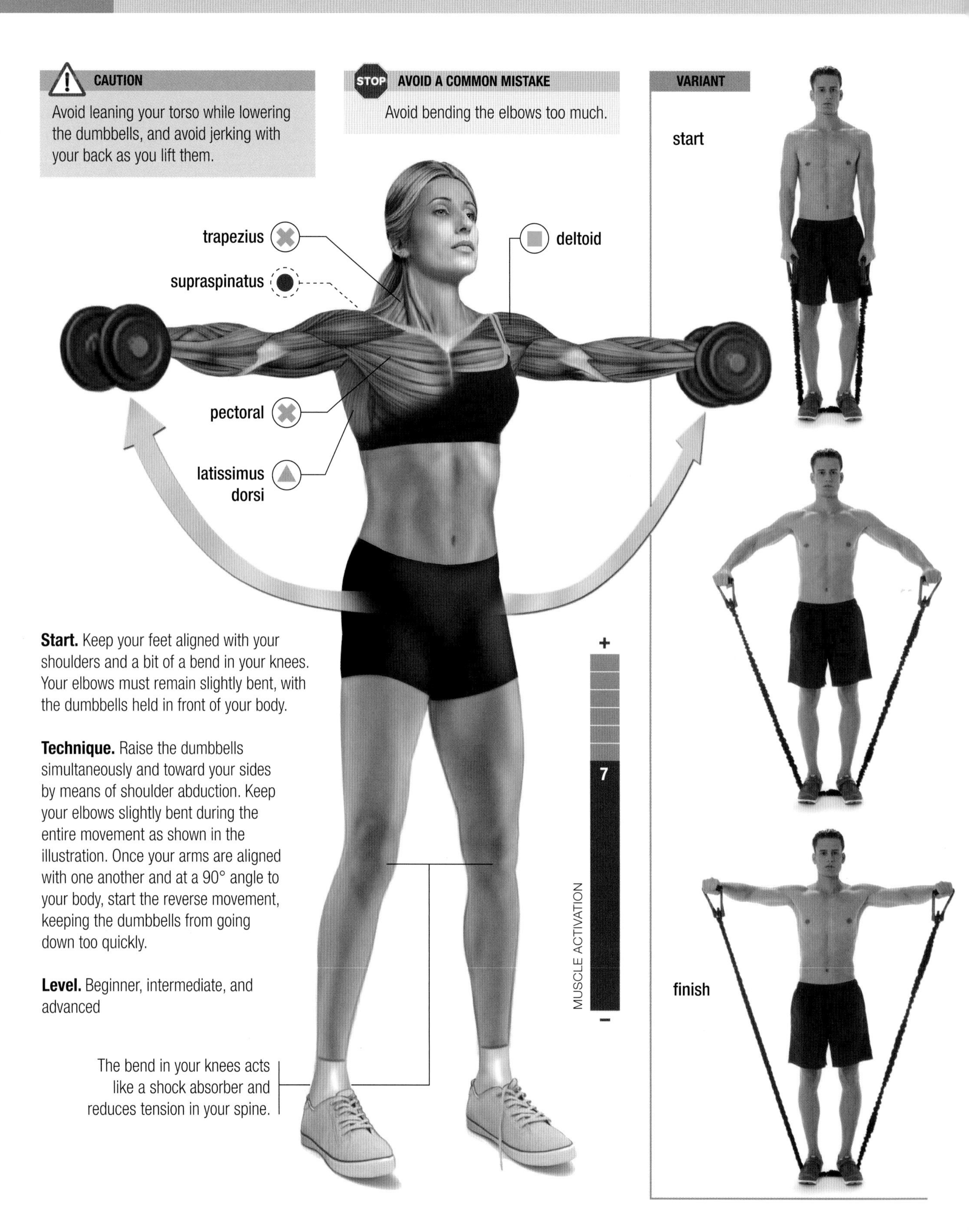

Start. Keep your feet aligned with your shoulders and a bit of a bend in your knees. Your elbows must remain slightly bent, with the dumbbells held in front of your body.

Technique. Raise the dumbbells simultaneously and toward your sides by means of shoulder abduction. Keep your elbows slightly bent during the entire movement as shown in the illustration. Once your arms are aligned with one another and at a 90° angle to your body, start the reverse movement, keeping the dumbbells from going down too quickly.

Level. Beginner, intermediate, and advanced

The bend in your knees acts like a shock absorber and reduces tension in your spine.

Deltoid

One-hand Lateral Raise

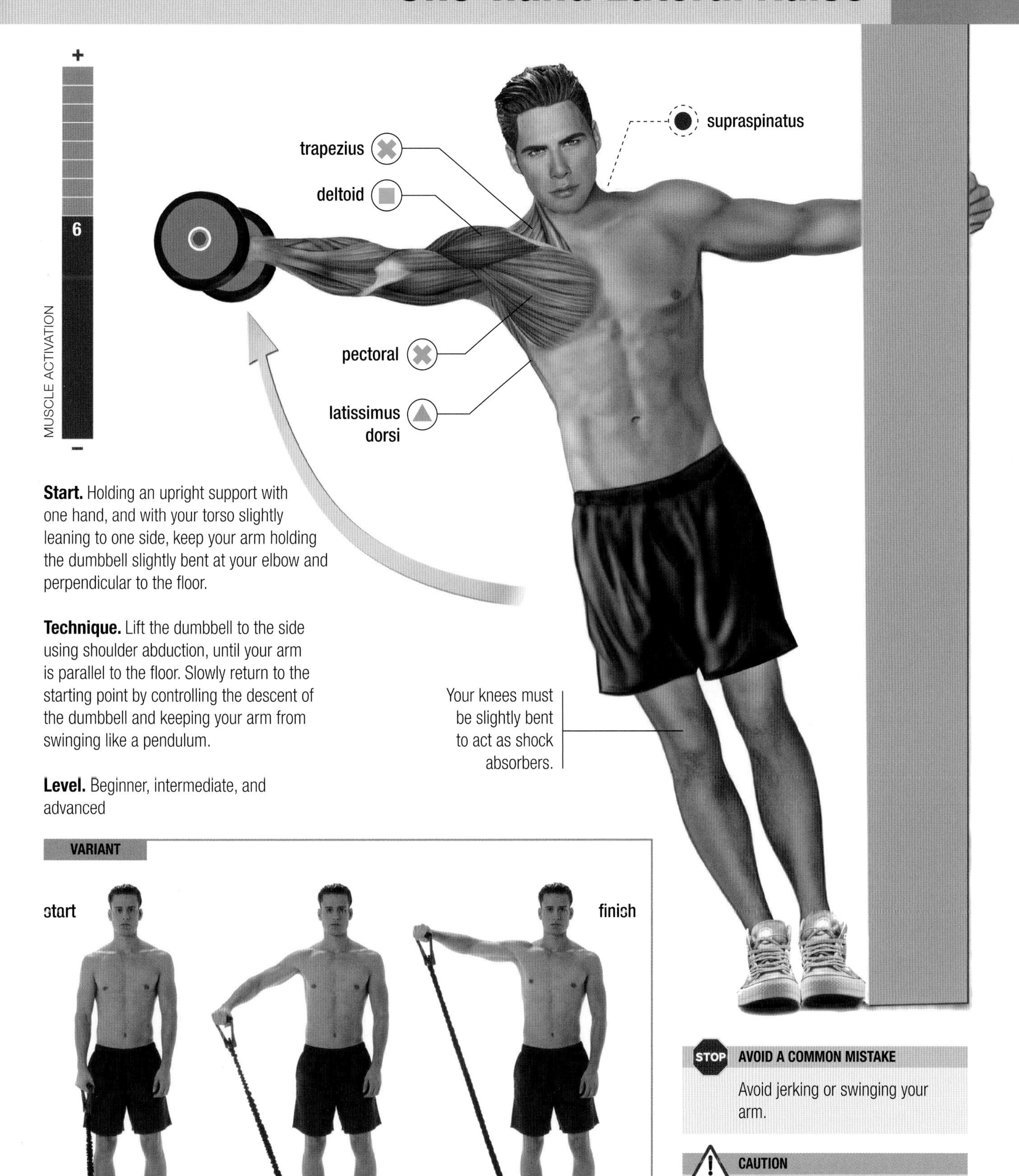

Start. Holding an upright support with one hand, and with your torso slightly leaning to one side, keep your arm holding the dumbbell slightly bent at your elbow and perpendicular to the floor.

Technique. Lift the dumbbell to the side using shoulder abduction, until your arm is parallel to the floor. Slowly return to the starting point by controlling the descent of the dumbbell and keeping your arm from swinging like a pendulum.

Level. Beginner, intermediate, and advanced

AVOID A COMMON MISTAKE

Avoid jerking or swinging your arm.

CAUTION

Make sure you are anchored to a stable support and maintain stability.

Deltoid

Lateral Raises on the Machine

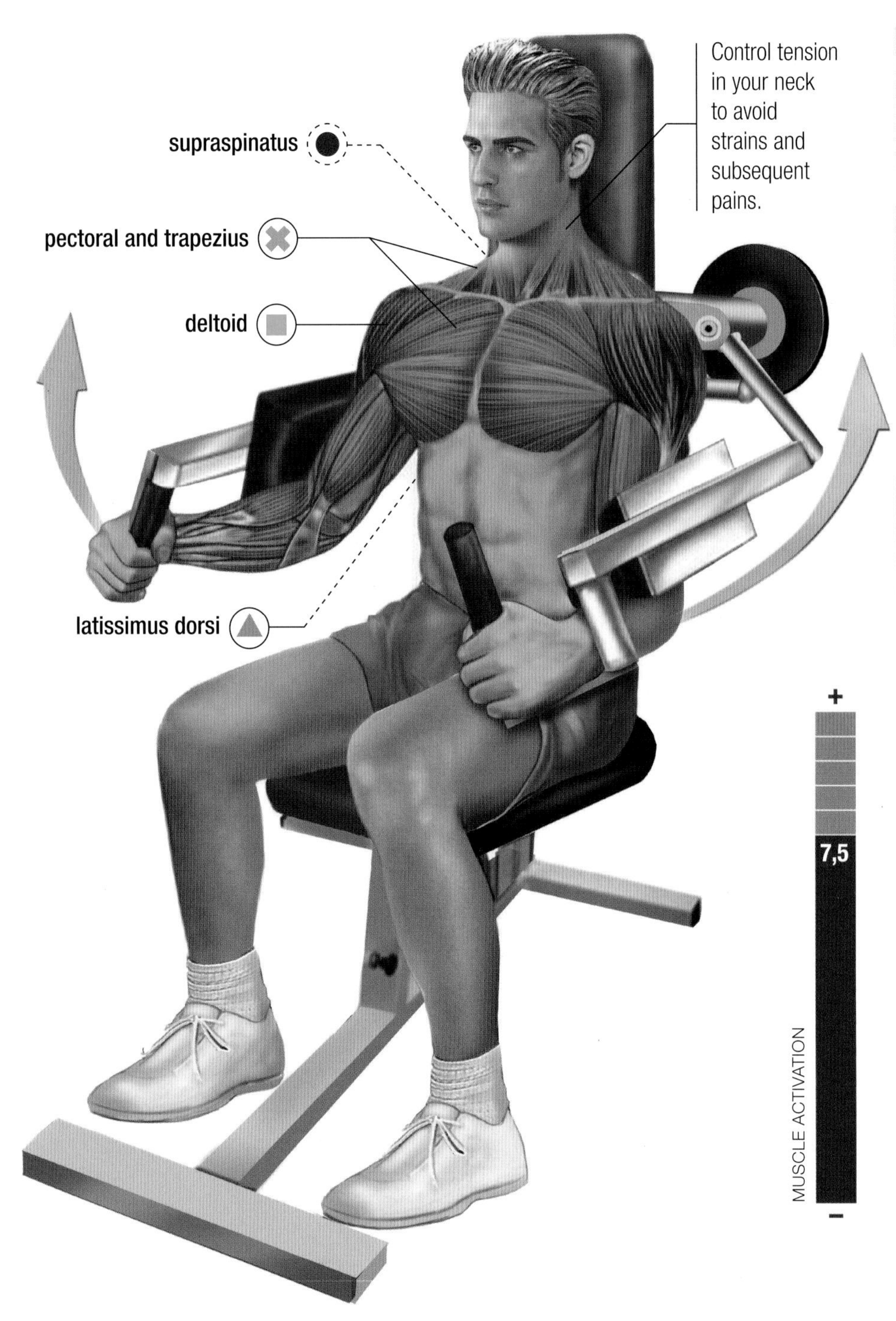

AVOID A COMMON MISTAKE

Avoid major contraction of the trapezius, because the force must be concentrated in the deltoids.

CAUTION

Make sure that the weight falls on the elbow cushions and not on your hands.

VARIANT

Start. Sit down with the outside of your elbows resting against the cushions. Remember to keep your back pressed tightly against the backrest on the machine, and that the hand grips are only a holding point, not the place where you exert your strength.

Technique. Lift your elbows using shoulder abduction, until your arms form a 90° angle with your torso. Lower your arms slowly so you can control the movement.

Level. Beginner, intermediate, and advanced

Reclining Lateral Raises

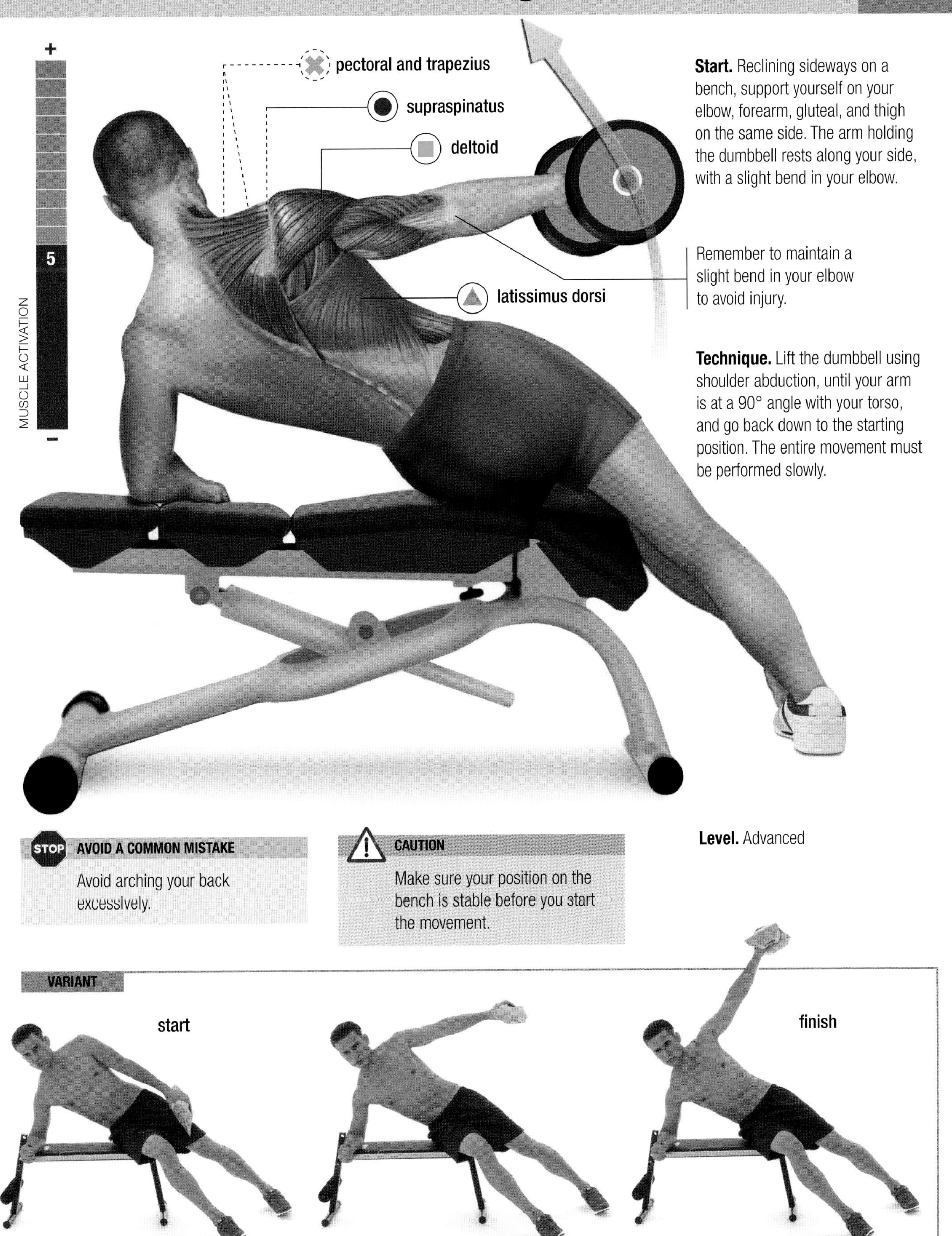

Start. Reclining sideways on a bench, support yourself on your elbow, forearm, gluteal, and thigh on the same side. The arm holding the dumbbell rests along your side, with a slight bend in your elbow.

Remember to maintain a slight bend in your elbow to avoid injury.

Technique. Lift the dumbbell using shoulder abduction, until your arm is at a 90° angle with your torso, and go back down to the starting position. The entire movement must be performed slowly.

Level. Advanced

AVOID A COMMON MISTAKE

Avoid arching your back excessively.

CAUTION

Make sure your position on the bench is stable before you start the movement.

Deltoid

Front Dumbbell Raises

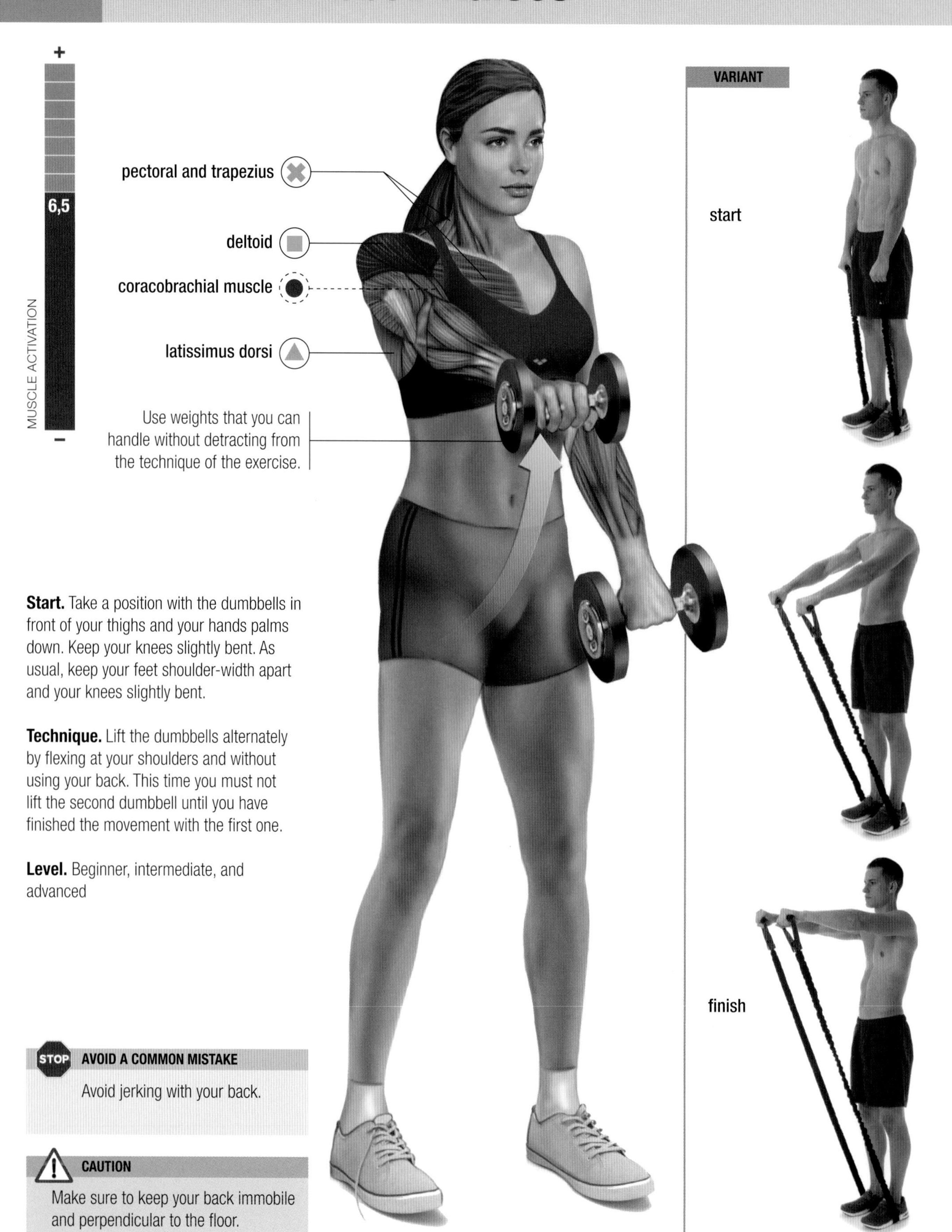

Start. Take a position with the dumbbells in front of your thighs and your hands palms down. Keep your knees slightly bent. As usual, keep your feet shoulder-width apart and your knees slightly bent.

Technique. Lift the dumbbells alternately by flexing at your shoulders and without using your back. This time you must not lift the second dumbbell until you have finished the movement with the first one.

Level. Beginner, intermediate, and advanced

AVOID A COMMON MISTAKE

Avoid jerking with your back.

CAUTION

Make sure to keep your back immobile and perpendicular to the floor.

Deltoid

Military Press

VARIANT

Start. Sit down on a bench with a backrest and hold the bar slightly to the front and above your collarbones.

Technique. Lift the bar using shoulder abduction and by extending your elbows until they are straight. Then lower the bar at a controlled speed without touching your chest.

Level. Intermediate and advanced

AVOID A COMMON MISTAKE

Don't restrict yourself to a very short movement. Perform the entire movement, even though this means working with a lighter weight.

CAUTION

Keep your back pressed against the backrest and avoid lowering your arms to an angle less than 45° with your trunk.

Deltoid

Arnold Press

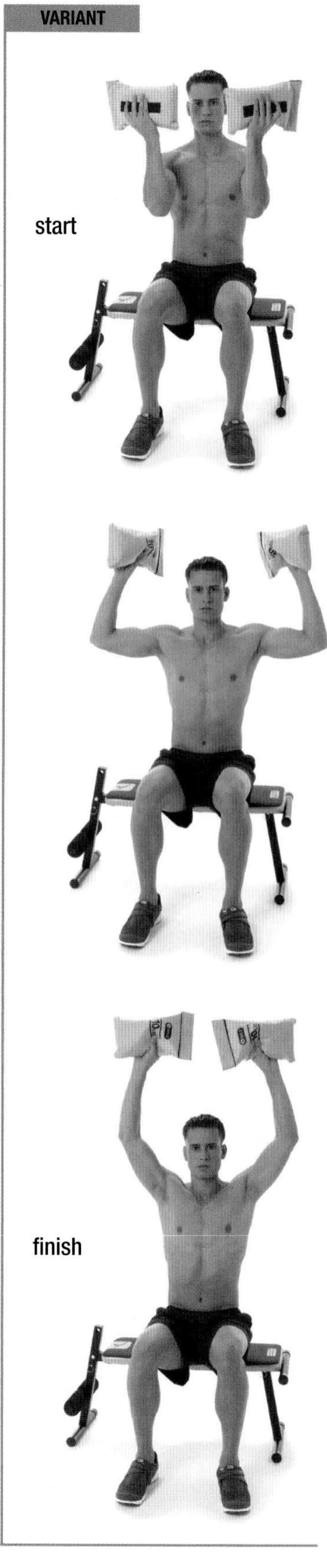

Start. Sit on a bench, your elbows bent in front of your body and holding a pair of dumbbells, with palms facing you.

Technique. Raise the dumbbells together and rotate your wrists so that, at the top of the motion, your palms end up facing forward. Your elbows should straighten and move from the front to the sides and upward. Then perform the reverse movement, returning to the starting position.

Level. Advanced

STOP **AVOID A COMMON MISTAKE**

Avoid arching your back.

CAUTION

Use a bench with a backrest.

Seated Dumbbell Press

Start. Sit on a bench with a backrest, or on a Scott bench, pressing your back against the cushion. Hold your arms to the sides, forming a 45° angle with your trunk and with your elbows bent. Your palms should face forward.

Technique. Raise the dumbbells and move them together at the top of the movement. Then bend your elbows and return to the starting point, but without lowering your arms too much. Maintain slow speed and good control over the technical movement throughout.

Level. Intermediate and advanced

AVOID A COMMON MISTAKE

Avoid arching your back beyond its natural curvature.

CAUTION

Always use a bench with a backrest in doing this exercise.

Deltoid

Shoulder Press on the Machine

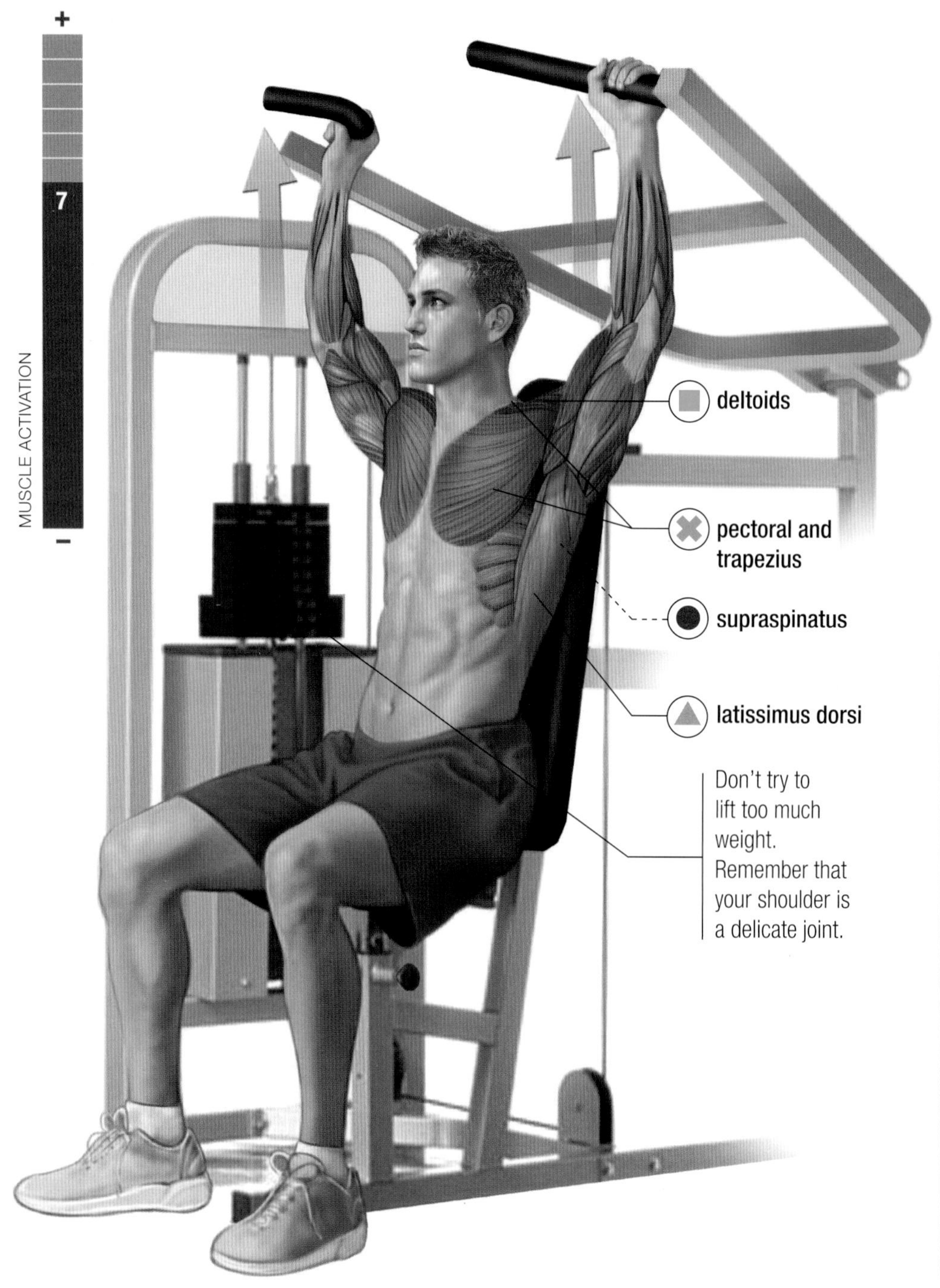
+
7
MUSCLE ACTIVATION
–
deltoids
pectoral and trapezius
supraspinatus
latissimus dorsi
Don't try to lift too much weight. Remember that your shoulder is a delicate joint.

VARIANT

start
finish

Start. Sit at the machine with your back against the backrest. Your arms are to your sides at a 45° angle with your trunk and with your elbows bent. Normally, these machines have two grips, one for pronation and another one that's neutral. You can use either one, keeping in mind that the pronation grip affects the middle of the deltoid more and the neutral one affects the anterior deltoid.

Technique. Straighten your elbows and lift against the resistance. The movement is guided by the machine, so there is not much chance for error. Once you reach the top, slowly lower the weight.

Level. Beginner, intermediate, and advanced

STOP

AVOID A COMMON MISTAKE

Avoid arching your back too much.

CAUTION

The angle that your arm forms with your trunk should be no smaller than 45° at any time. This will reduce the risk of injury to your shoulder.

Posterior Deltoids on the Machine

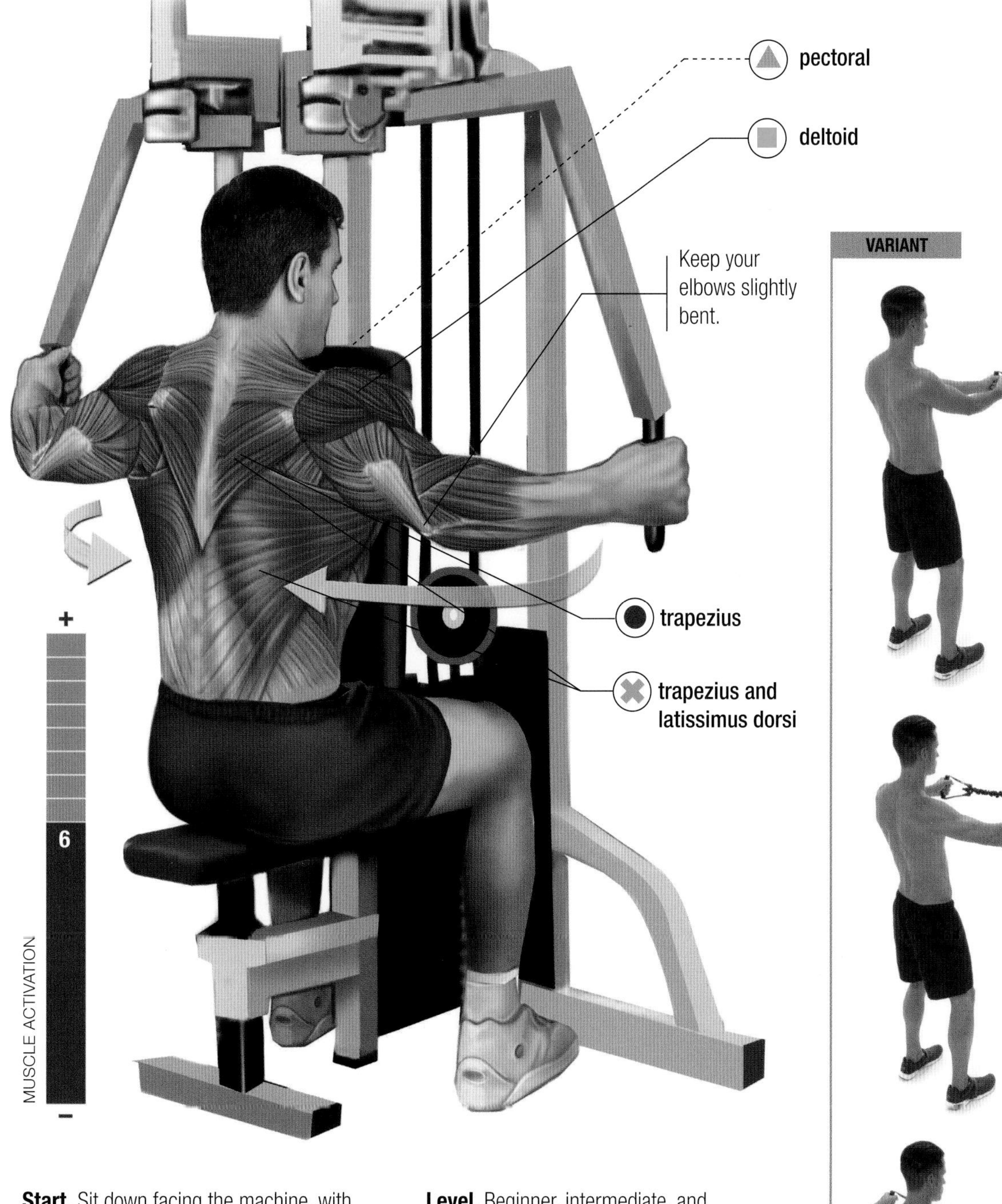

Start. Sit down facing the machine, with your chest pressed against the cushion.

Technique. Extend your shoulders by drawing your hands toward the rear, without breaking contact between your chest and the cushion. Once you reach the point of maximum extension, perform the inverse movement and return to the starting point. You can also do this exercise seated facing the opposite direction on a peck-deck machine and pressing against the cushions with your elbows.

Level. Beginner, intermediate, and advanced

STOP **AVOID A COMMON MISTAKE**

Don't break contact between your chest and the cushion.

 **CAUTION**

Adjust the height of the seat so that the grips are the same height as your shoulders or slightly lower.

Deltoid

Seated Posterior Deltoids or Flies

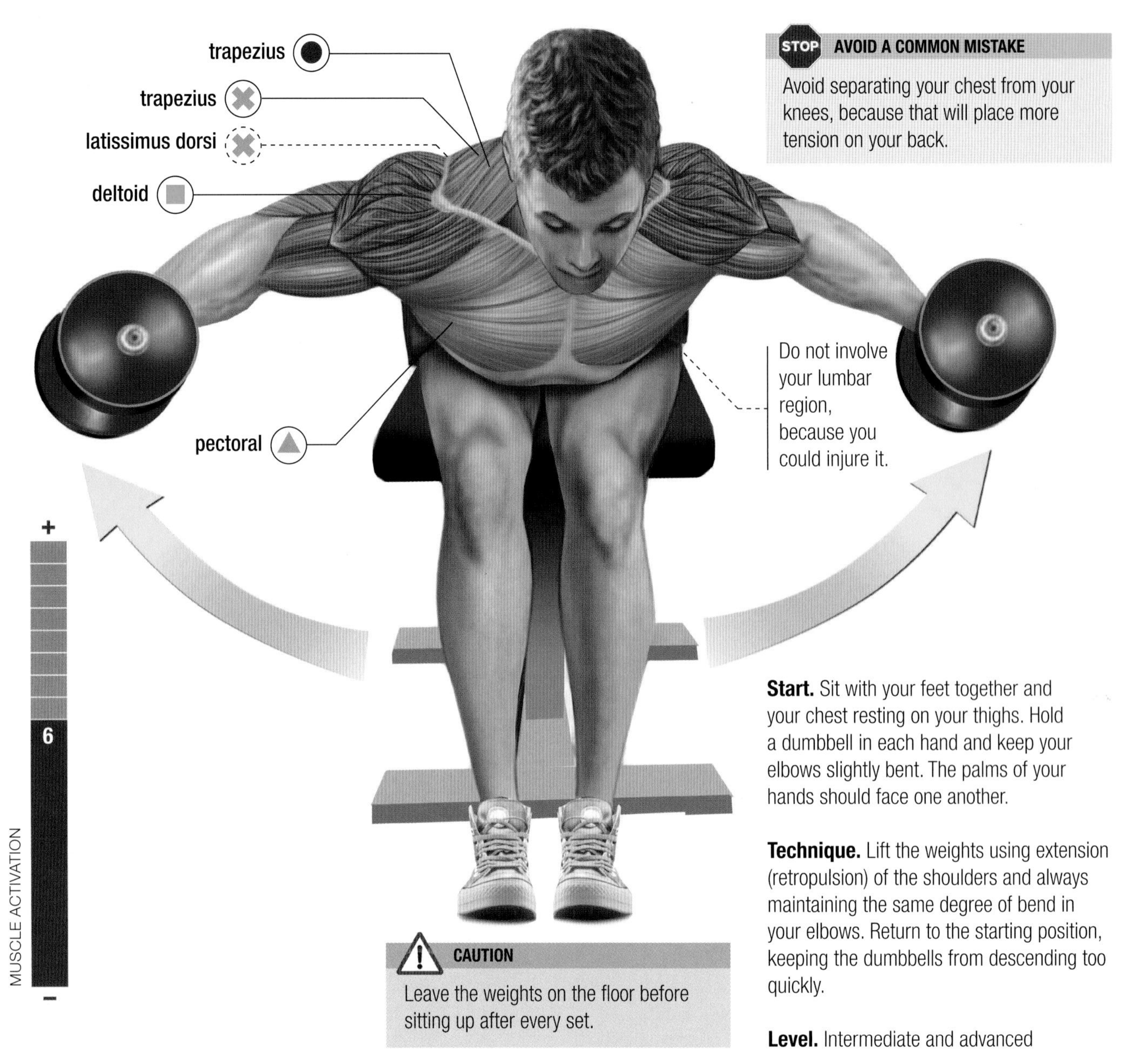

STOP AVOID A COMMON MISTAKE

Avoid separating your chest from your knees, because that will place more tension on your back.

CAUTION

Leave the weights on the floor before sitting up after every set.

Start. Sit with your feet together and your chest resting on your thighs. Hold a dumbbell in each hand and keep your elbows slightly bent. The palms of your hands should face one another.

Technique. Lift the weights using extension (retropulsion) of the shoulders and always maintaining the same degree of bend in your elbows. Return to the starting position, keeping the dumbbells from descending too quickly.

Level. Intermediate and advanced

VARIANT

One-hand Posterior Deltoids

STOP AVOID A COMMON MISTAKE

Avoid pivoting your upper body in order to move the weight.

CAUTION

Rest your free hand on your thigh to help support your back.

Keep your back straight and your gluteals to the outside, to reduce the risk of injury.

Start. Stand with your feet apart, your knees at 135°, and your upper body leaning forward without arching your back. Using the hand on the other side of your body from the machine, hold the handle with a neutral grip and your elbow nearly straight. The cable is set at the bottom of the machine.

Technique. As you extend your shoulder, raise and move your hand away from the machine by making an arc. Both the upward and downward movements must be performed slowly.

Level. Advanced

VARIANT

Arms

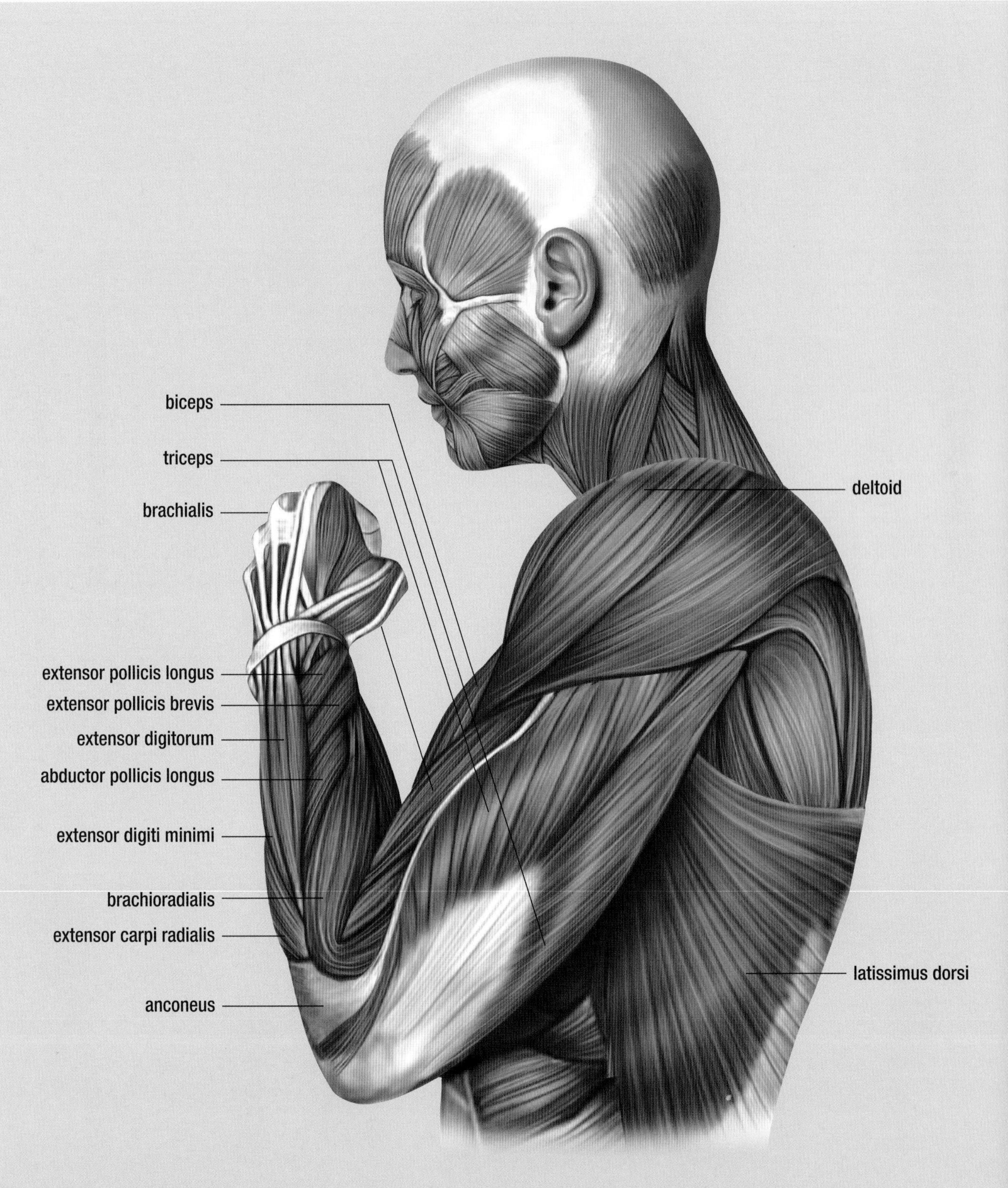
biceps
triceps
brachialis
extensor pollicis longus
extensor pollicis brevis
extensor digitorum
abductor pollicis longus
extensor digiti minimi
brachioradialis
extensor carpi radialis
anconeus
deltoid
latissimus dorsi

The upper arm is the part of your upper extremity that lies between the shoulder and the elbow; but, in this section, the book will also present the muscles of the forearm. In conjunction with the chest, the arm probably constitutes one of the main areas of concentration for bodybuilding athletes. There are two powerful, visible muscles in the upper arm: the biceps, which is located in the frontal area, and the triceps, which is in the rear.

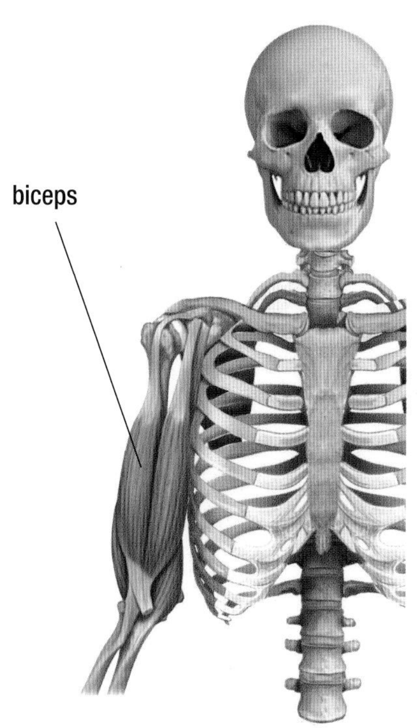

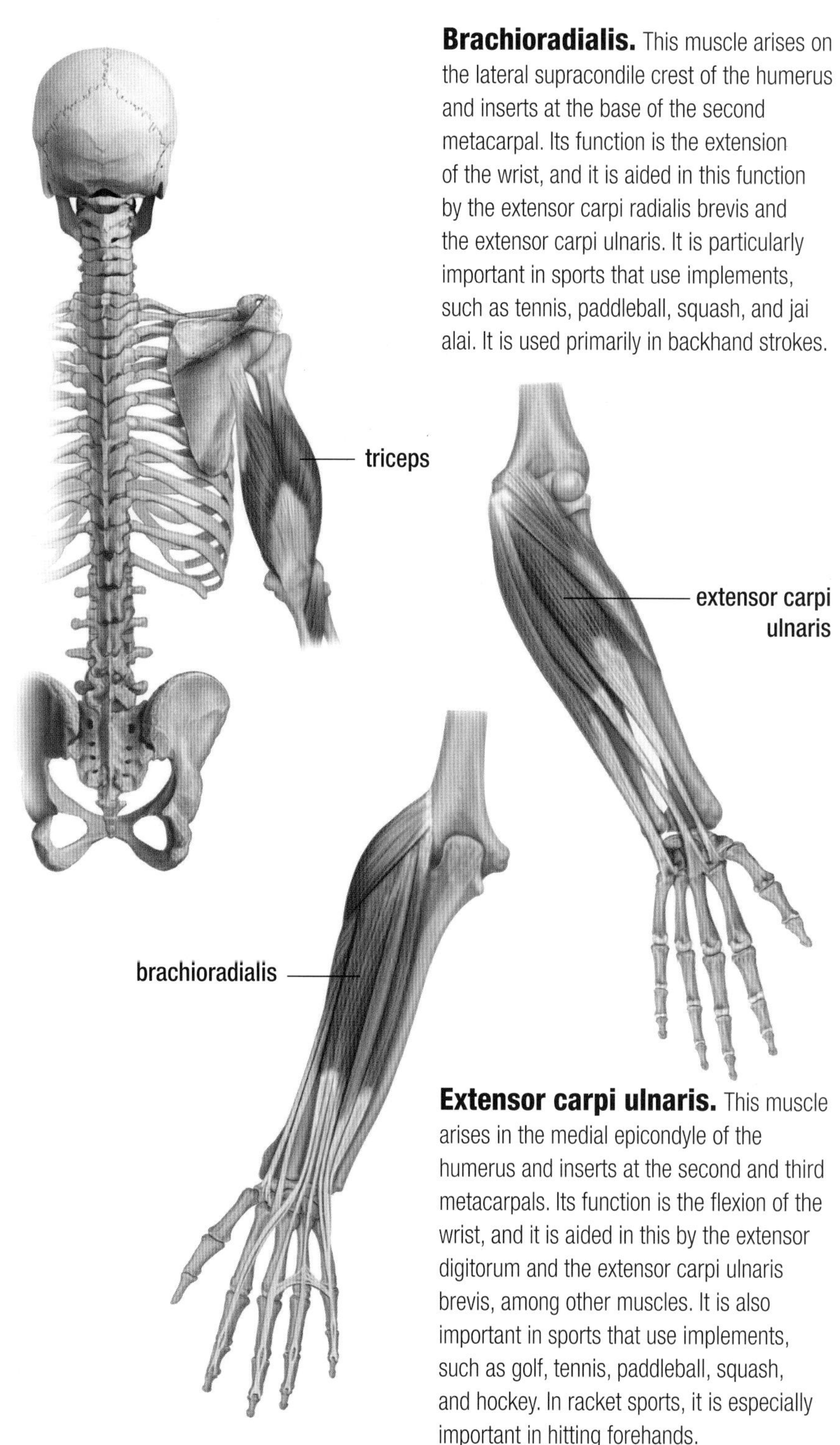

Biceps. This muscle has two parts. The first originates at the coracoid apophysis located on the scapula, and it inserts at the radial tuberosity (third proximal of the radius). The second one originates at the supraglenoid tubercle located on the scapula, and it inserts at the bicipital aponeurosis. The contraction of the biceps allows the bending of the elbow and the outward rotation of the wrist, and it is essential for hitting hard shots in racket sports, in climbing, and holding a ball against the torso—for example, in football.

Triceps. This muscle has three parts that arise at the infraglenoid tuberosity (scapula) and at the diaphysis of the humerus. The three parts have a common insertion at the olecranon, which is located on the ulna. The function of the triceps is to straighten the elbow; this muscle is vitally important in throwing, including basketball passes and handball, and in direct hits or jabs as in boxing.

Brachioradialis. This muscle arises on the lateral supracondile crest of the humerus and inserts at the base of the second metacarpal. Its function is the extension of the wrist, and it is aided in this function by the extensor carpi radialis brevis and the extensor carpi ulnaris. It is particularly important in sports that use implements, such as tennis, paddleball, squash, and jai alai. It is used primarily in backhand strokes.

Extensor carpi ulnaris. This muscle arises in the medial epicondyle of the humerus and inserts at the second and third metacarpals. Its function is the flexion of the wrist, and it is aided in this by the extensor digitorum and the extensor carpi ulnaris brevis, among other muscles. It is also important in sports that use implements, such as golf, tennis, paddleball, squash, and hockey. In racket sports, it is especially important in hitting forehands.

Biceps

Standing Barbell Curls

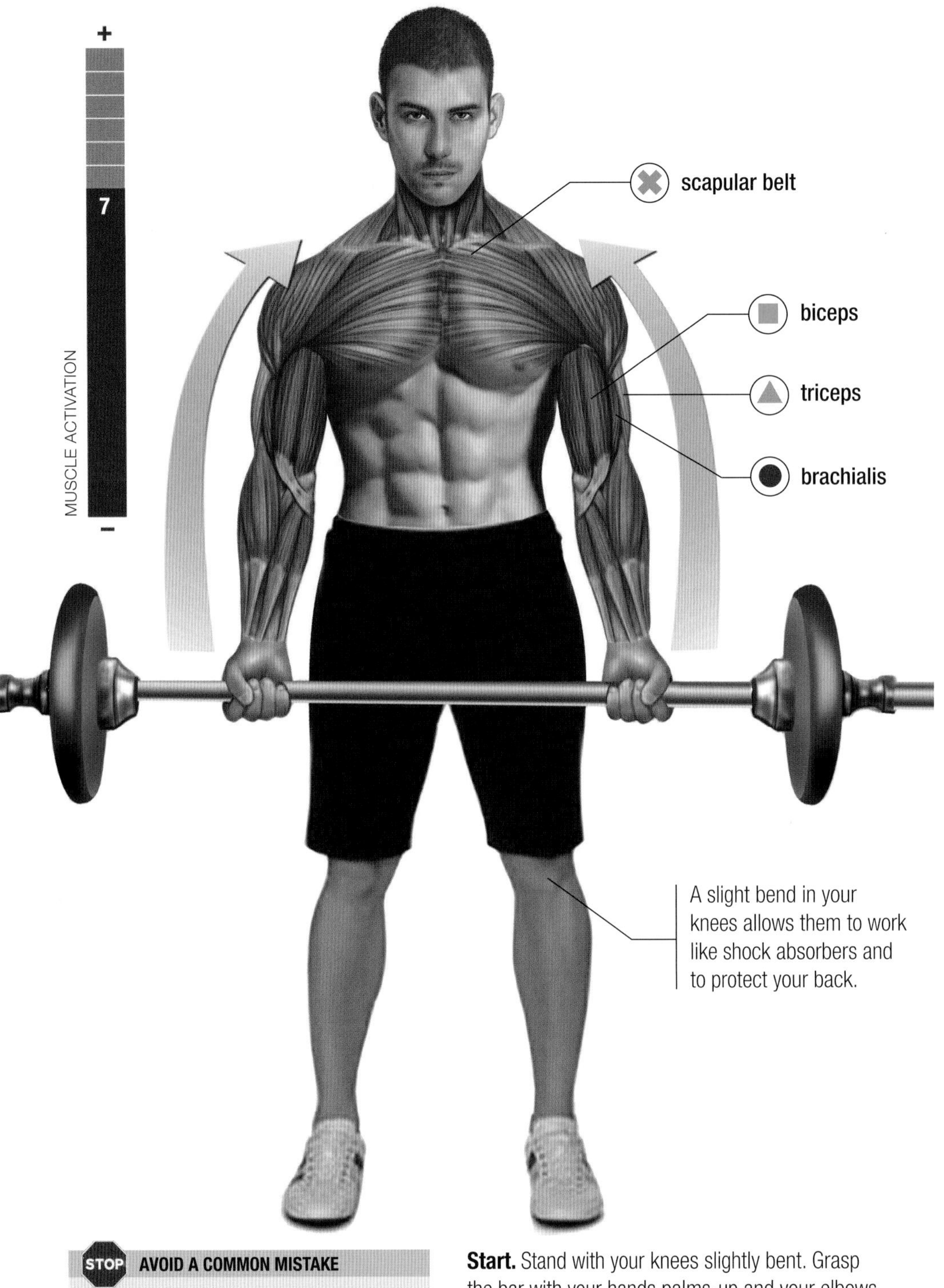
+
7
MUSCLE ACTIVATION
–
scapular belt
biceps
triceps
brachialis
A slight bend in your knees allows them to work like shock absorbers and to protect your back.

VARIANT
start
finish

AVOID A COMMON MISTAKE

Avoid jerking, using your back, or rocking.

CAUTION

Be sure to keep your feet wide enough apart to provide a good support base.

Start. Stand with your knees slightly bent. Grasp the bar with your hands palms-up and your elbows straight but not locked.

Technique. Lift the bar by bending at your elbows, keeping your upper arms anchored on your body, and without moving your elbows forward. It is important to avoid both rocking motions and using your shoulders during the movement.

Level. Beginner, intermediate, and advanced

Alternating Dumbbell Curls

+
6,5
MUSCLE ACTIVATION
–
Keep your upper arms anchored on your body and remember that the movement comes from your forearm.
scapular belt
biceps
triceps
brachialis
VARIANT
start
finish

Start. Stand with your elbows straight and hold the dumbbells with your palms facing your thighs.

Technique. Alternately raise the dumbbells by bending at your elbows. As one dumbbell goes down, the other should go up. There is no need to complete the movement with one to begin with the other. The overall movement must be slow and deliberate.

Level. Beginner, intermediate, and advanced

AVOID A COMMON MISTAKE

Avoid leaning your back, using rocking motions, or moving your shoulders.

CAUTION

Keep your back straight, your feet aligned with your shoulders, and a slight bend in your knees to achieve proper balance and avoid unnecessary tension in the lumbar region.

Biceps

Incline Curls

Start. Take a position on a 45° inclined bench, with your elbows straight and your back supported by the backrest on the bench. Your feet should be slightly separated.

Technique. Bend at your elbows and raise the weights without using your deltoids.

Level. Intermediate and advanced

AVOID A COMMON MISTAKE

Avoid jerking or swinging your hands like a pendulum.

CAUTION

In the negative phase, do not let the weight drop, because this could cause excess tension in your shoulders.

VARIANT

Scott Curls

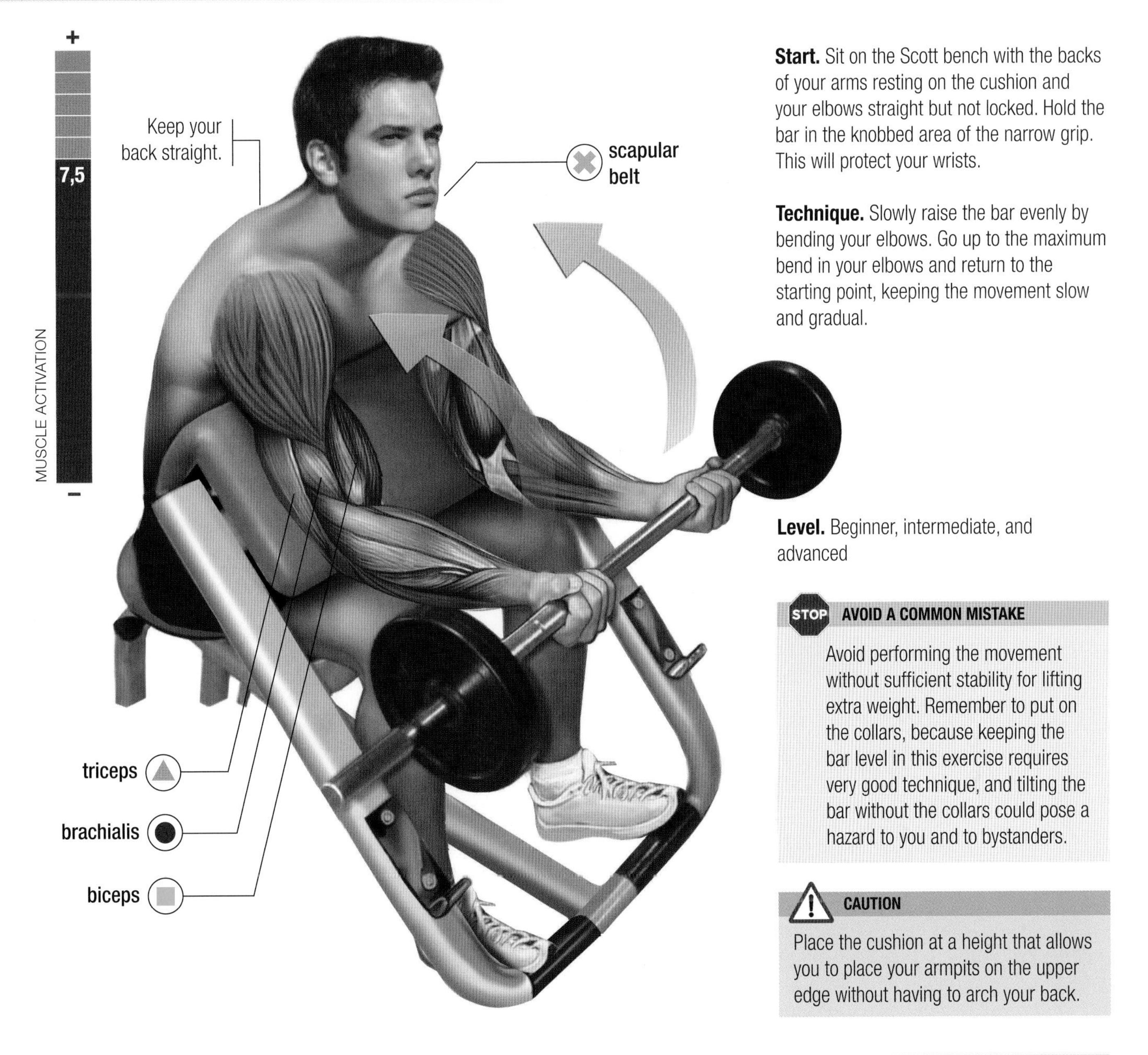

Start. Sit on the Scott bench with the backs of your arms resting on the cushion and your elbows straight but not locked. Hold the bar in the knobbed area of the narrow grip. This will protect your wrists.

Technique. Slowly raise the bar evenly by bending your elbows. Go up to the maximum bend in your elbows and return to the starting point, keeping the movement slow and gradual.

Level. Beginner, intermediate, and advanced

STOP AVOID A COMMON MISTAKE

Avoid performing the movement without sufficient stability for lifting extra weight. Remember to put on the collars, because keeping the bar level in this exercise requires very good technique, and tilting the bar without the collars could pose a hazard to you and to bystanders.

CAUTION

Place the cushion at a height that allows you to place your armpits on the upper edge without having to arch your back.

VARIANT

Biceps

Concentration Curls

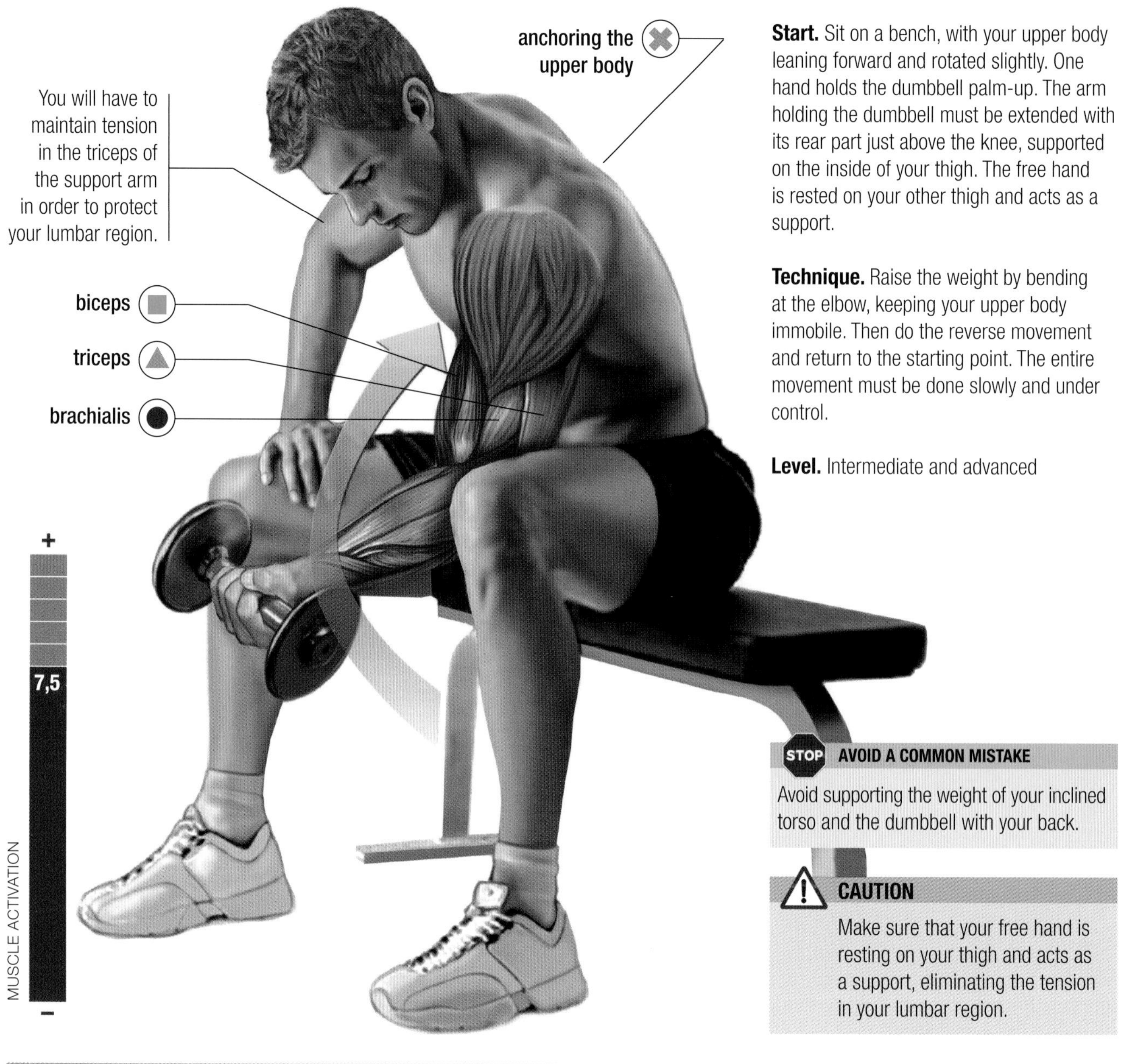

Start. Sit on a bench, with your upper body leaning forward and rotated slightly. One hand holds the dumbbell palm-up. The arm holding the dumbbell must be extended with its rear part just above the knee, supported on the inside of your thigh. The free hand is rested on your other thigh and acts as a support.

Technique. Raise the weight by bending at the elbow, keeping your upper body immobile. Then do the reverse movement and return to the starting point. The entire movement must be done slowly and under control.

Level. Intermediate and advanced

STOP AVOID A COMMON MISTAKE

Avoid supporting the weight of your inclined torso and the dumbbell with your back.

CAUTION

Make sure that your free hand is resting on your thigh and acts as a support, eliminating the tension in your lumbar region.

VARIANT

Hammer Curls

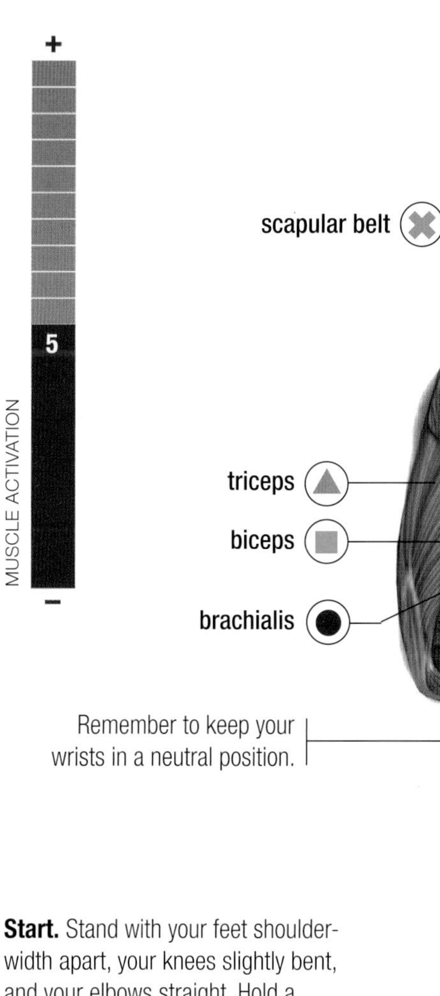

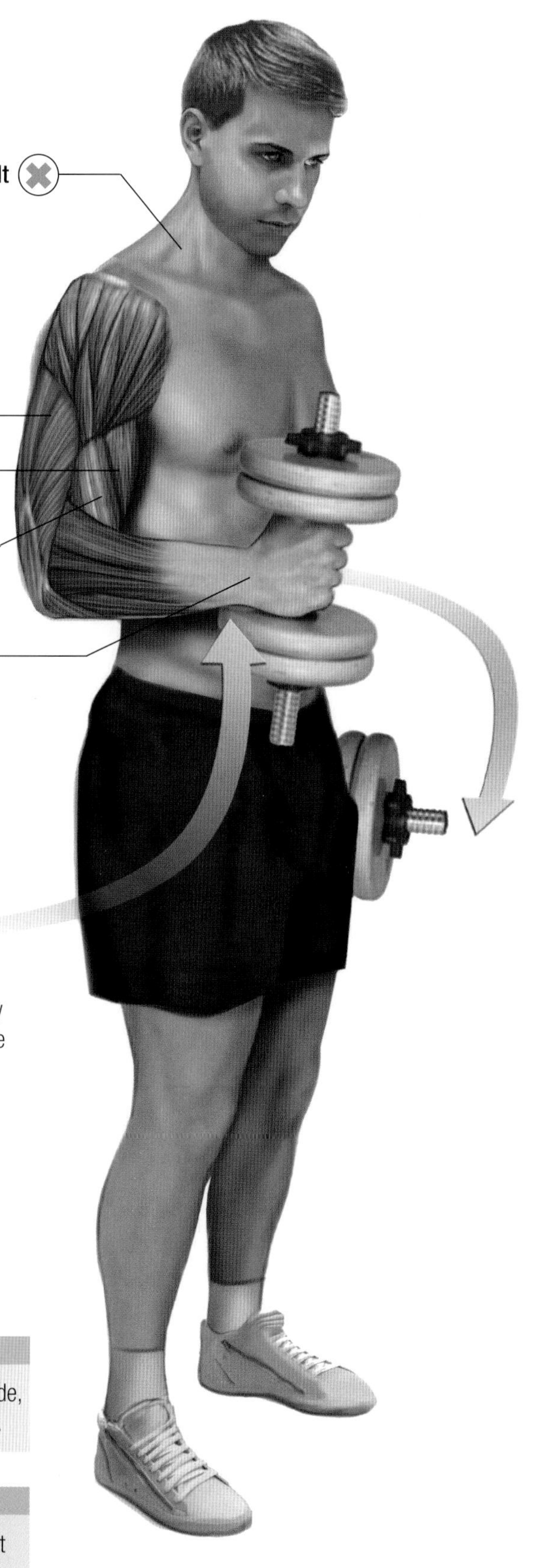

Remember to keep your wrists in a neutral position.

Start. Stand with your feet shoulder-width apart, your knees slightly bent, and your elbows straight. Hold a dumbbell in each hand, with your palms facing inward.

Technique. Lift the dumbbells alternately by bending at your elbows. Straighten one elbow as you bend the other one, so that both arms are working at the same time but moving in opposite directions.

Level. Beginner, intermediate, and advanced

AVOID A COMMON MISTAKE

Avoid leaning your upper body to one side, according to the movement of the arms.

CAUTION

Keep the upper body still, without rocking.

VARIANT

Triceps

French Press

Start. Lie down on a bench and keep your elbows bent. Use a narrow grip as you hold an EZ-bar near your forehead.

Technique. Slowly straighten your elbows, keeping your arms perpendicular to the floor. Straighten your arms all the way and return to the starting point at a slow, controlled speed.

Level. Advanced

AVOID A COMMON MISTAKE

Don't move your elbows outward as you raise the bar.

CAUTION

Keep your arms perpendicular to the floor during the entire exercise.

VARIANT

start

finish

Regular Pull-down

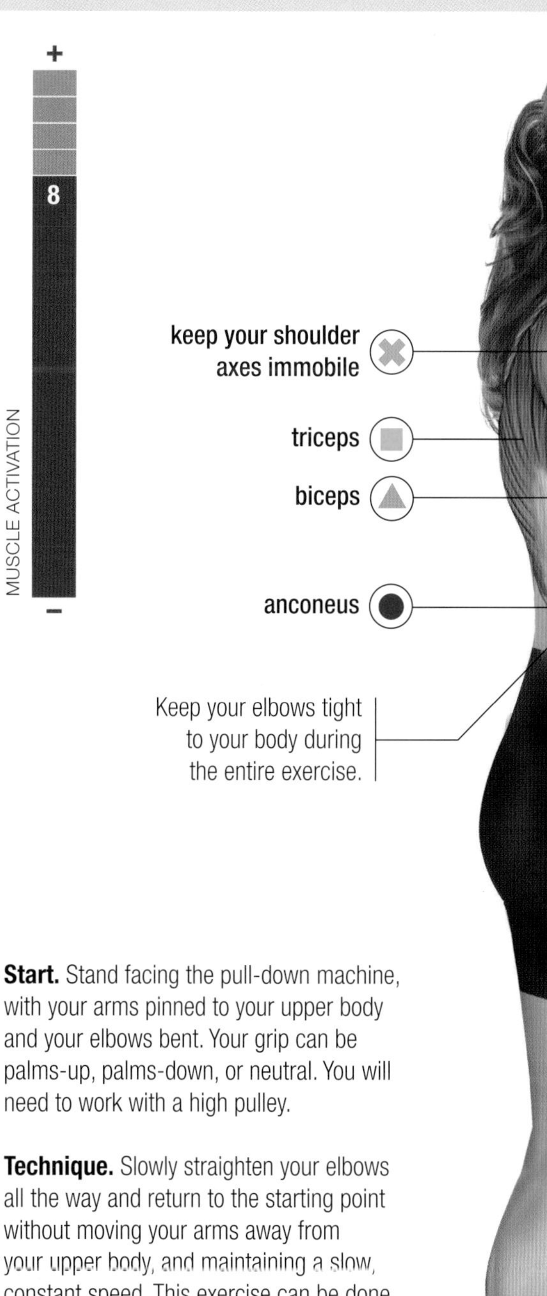

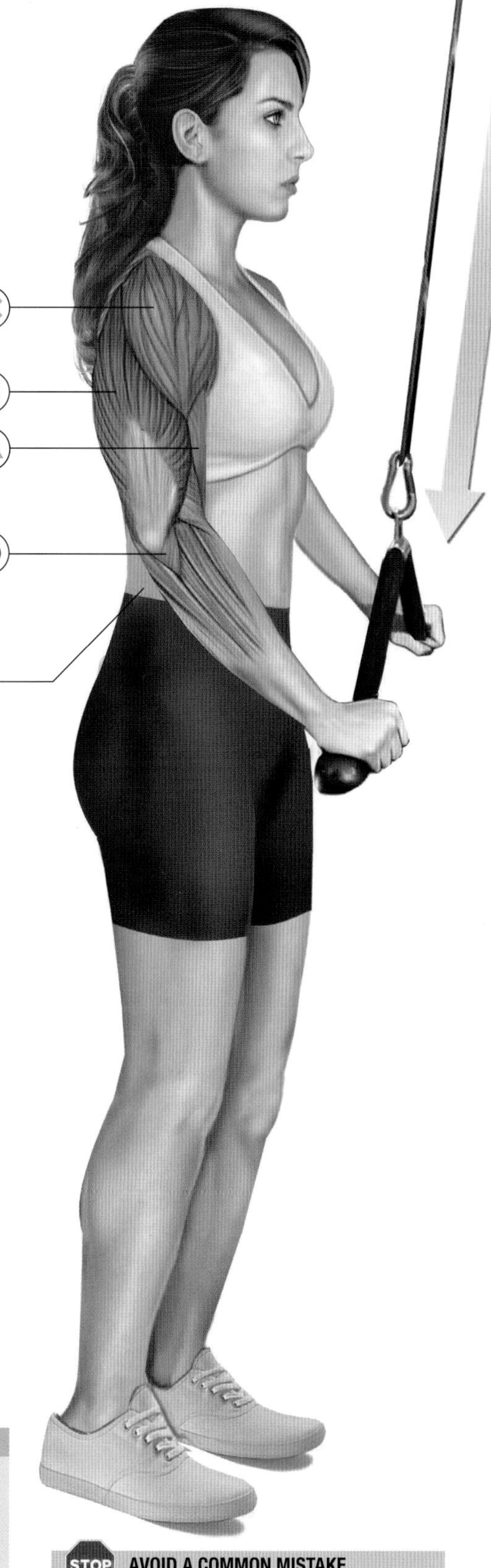

Start. Stand facing the pull-down machine, with your arms pinned to your upper body and your elbows bent. Your grip can be palms-up, palms-down, or neutral. You will need to work with a high pulley.

Technique. Slowly straighten your elbows all the way and return to the starting point without moving your arms away from your upper body, and maintaining a slow, constant speed. This exercise can be done with a rope grip attachment.

Level. Beginner, intermediate, and advanced

CAUTION

Maintain a good base of support, whether with one foot ahead of the other or with both feet aligned with your shoulders. This will encourage stability, even when working with heavier resistance.

STOP **AVOID A COMMON MISTAKE**

Avoid bending and extending your shoulders during the movement.

Triceps

Seated Dumbbell Press

Start. Sit on a Scott bench, pressing your back against the arm cushion. If you don't have a Scott bench, use a regular bench, but remember that it will not be as stable. Keep your arms near your head and your elbows bent, so that the dumbbell hangs down behind.

Technique. Straighten your elbows as you try to keep them in a fixed point in space, without moving your arms away from your head. Continue until your elbows are nearly totally straight, and then return to the starting point. This exercise can be done using two hands and just one dumbbell, or with one hand and one dumbbell.

Level. Intermediate and advanced

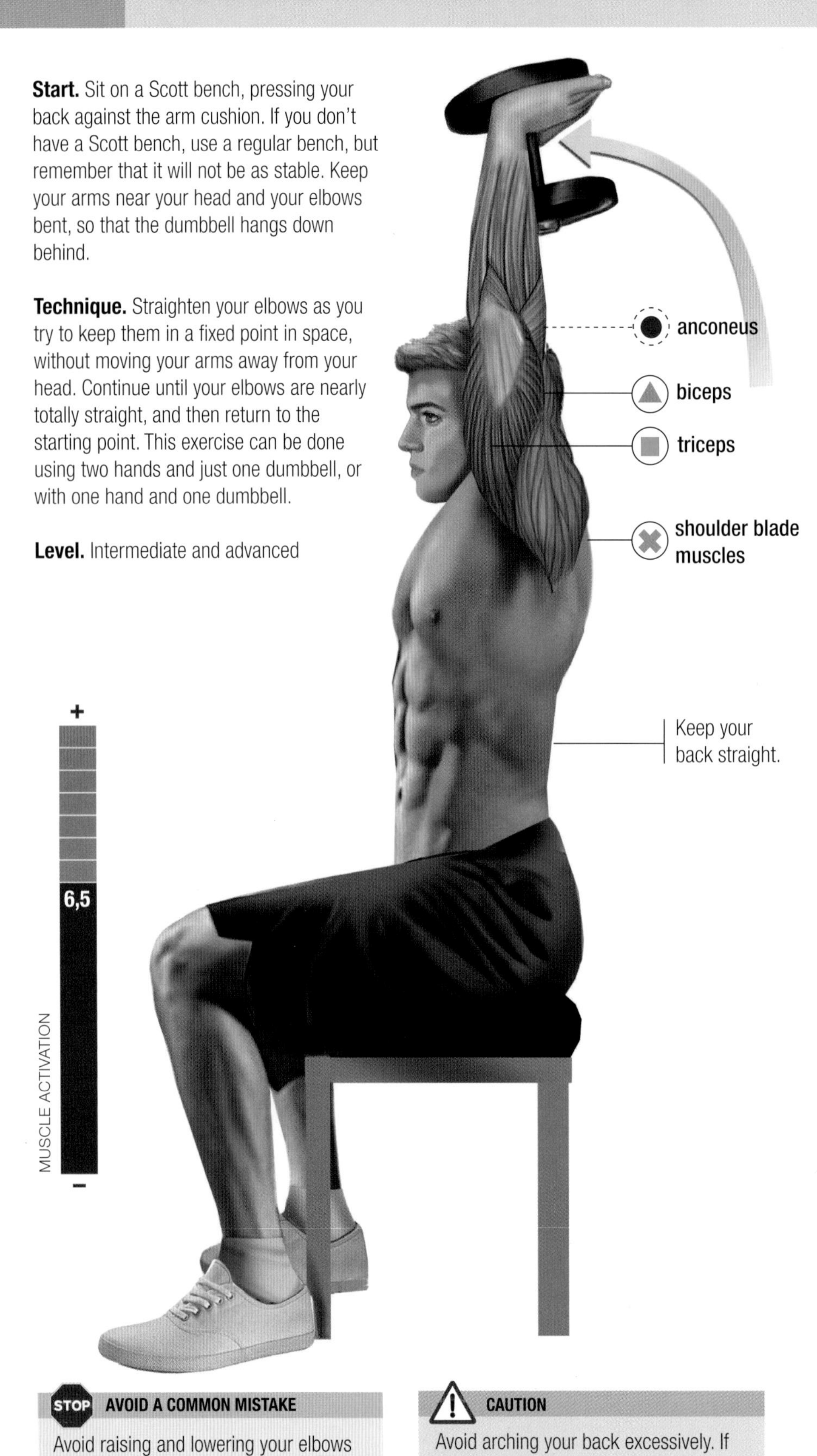

STOP AVOID A COMMON MISTAKE

Avoid raising and lowering your elbows during the performance of this exercise. Keep them close to your head.

CAUTION

Avoid arching your back excessively. If you have a Scott bench, use it.

Triceps Dips

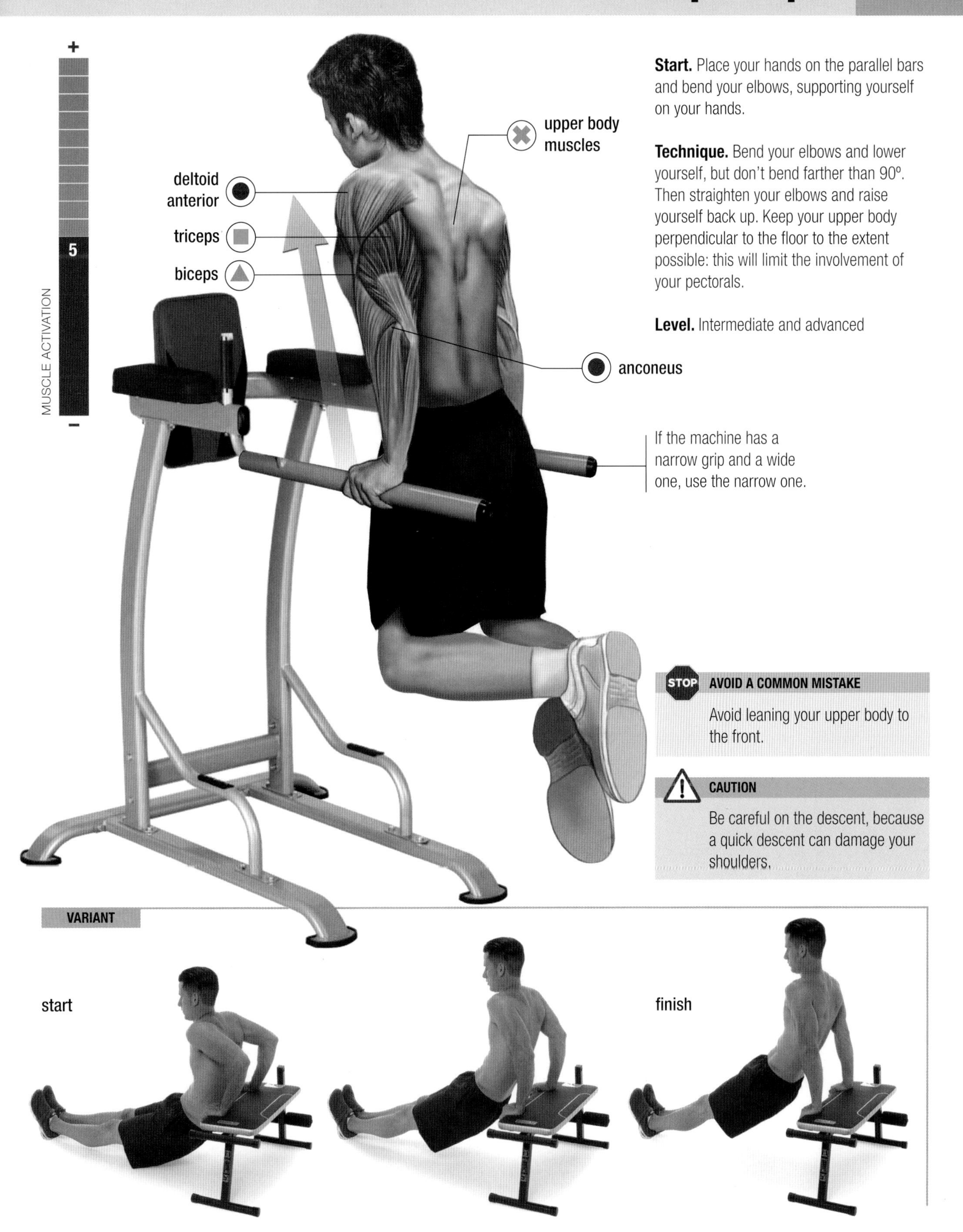

Start. Place your hands on the parallel bars and bend your elbows, supporting yourself on your hands.

Technique. Bend your elbows and lower yourself, but don't bend farther than 90º. Then straighten your elbows and raise yourself back up. Keep your upper body perpendicular to the floor to the extent possible: this will limit the involvement of your pectorals.

Level. Intermediate and advanced

If the machine has a narrow grip and a wide one, use the narrow one.

STOP AVOID A COMMON MISTAKE

Avoid leaning your upper body to the front.

CAUTION

Be careful on the descent, because a quick descent can damage your shoulders.

Triceps

Triceps Back Kick

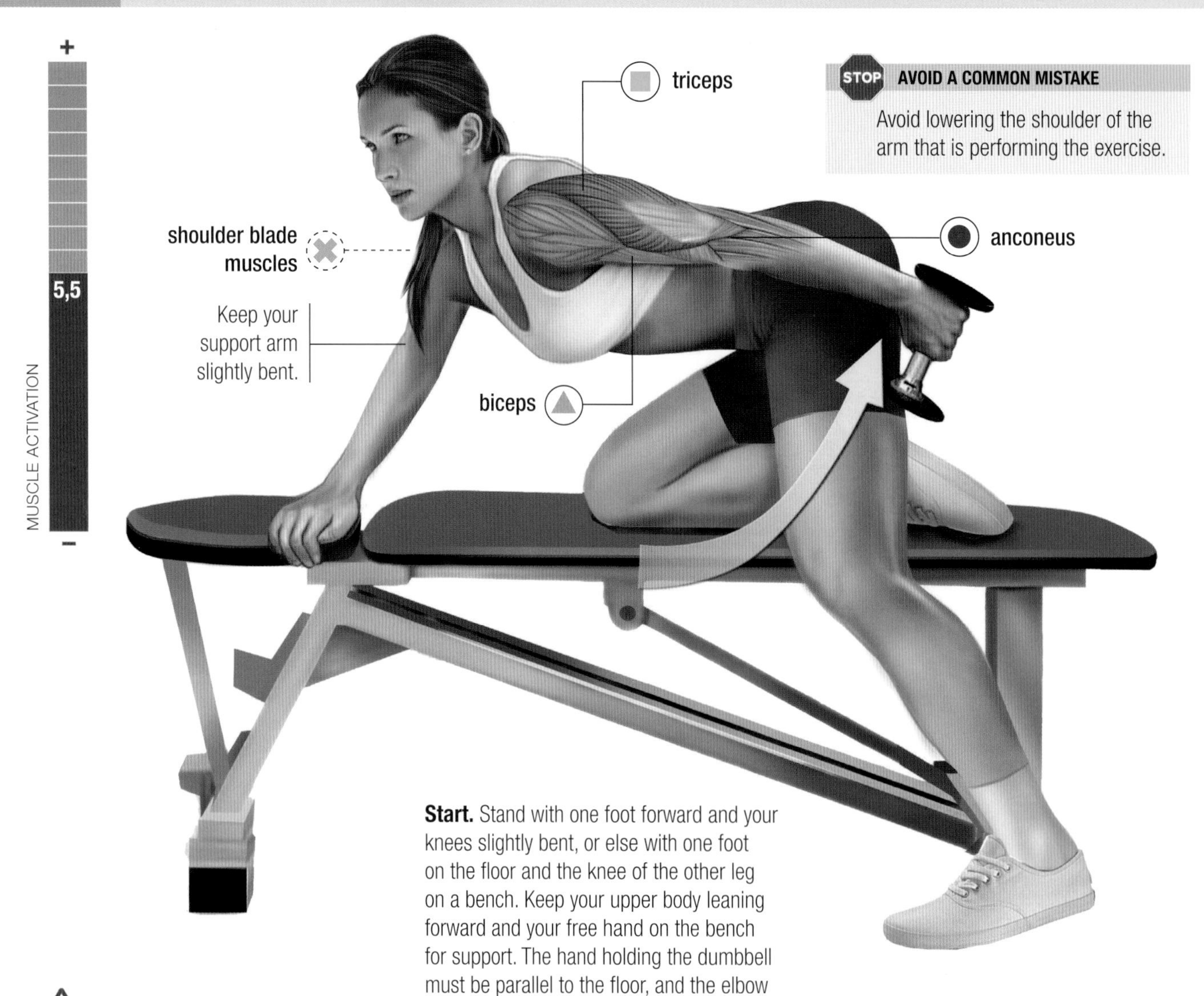

STOP **AVOID A COMMON MISTAKE**

Avoid lowering the shoulder of the arm that is performing the exercise.

Start. Stand with one foot forward and your knees slightly bent, or else with one foot on the floor and the knee of the other leg on a bench. Keep your upper body leaning forward and your free hand on the bench for support. The hand holding the dumbbell must be parallel to the floor, and the elbow must be bent.

Technique. Straighten your elbow without letting it move downward, keeping the upper arm parallel to the floor. Achieve complete extension and let the weight back down slowly, maintaining control over the movement throughout.

Level. Intermediate and advanced

CAUTION

Keep your free arm in contact with the bench to avoid excessive tension in your back.

VARIANT

start

finish

Narrow-grip Bench Press

+
6
MUSCLE ACTIVATION
–
Make sure you put the collars on the bar, since the narrow grip reduces its stability.
biceps
pectoral
triceps
shoulder blade muscles

Start. Lie down on a pressing bench. Hold the bar with both hands separated by about a hand's width. Keep your elbows straight.

Technique. Lift the bar from the supports and lower it to your chest by bending your elbows. When you reach the bottom, perform the reverse movement. Don't move your elbows outward during the motion.

Level. Advanced

STOP AVOID A COMMON MISTAKE

Avoid working with an excessively wide grip.

CAUTION

Keep your back pressed tightly to the bench, but maintain its natural curvature.

VARIANT

start

finish

Forearms

Back Barbell Wrist Curls

Start. Stand with your arms straight, holding the bar behind your body, and your palms facing rearward. Keep your feet in line with your shoulders.

Technique. Raise the bar slightly by flexing your wrists. You will see that the movement is very short, and that's the way it has to be. If you try to move the joint further than its normal range of motion, you will probably interfere with the technique of the exercise.

Level. Beginner, intermediate, and advanced

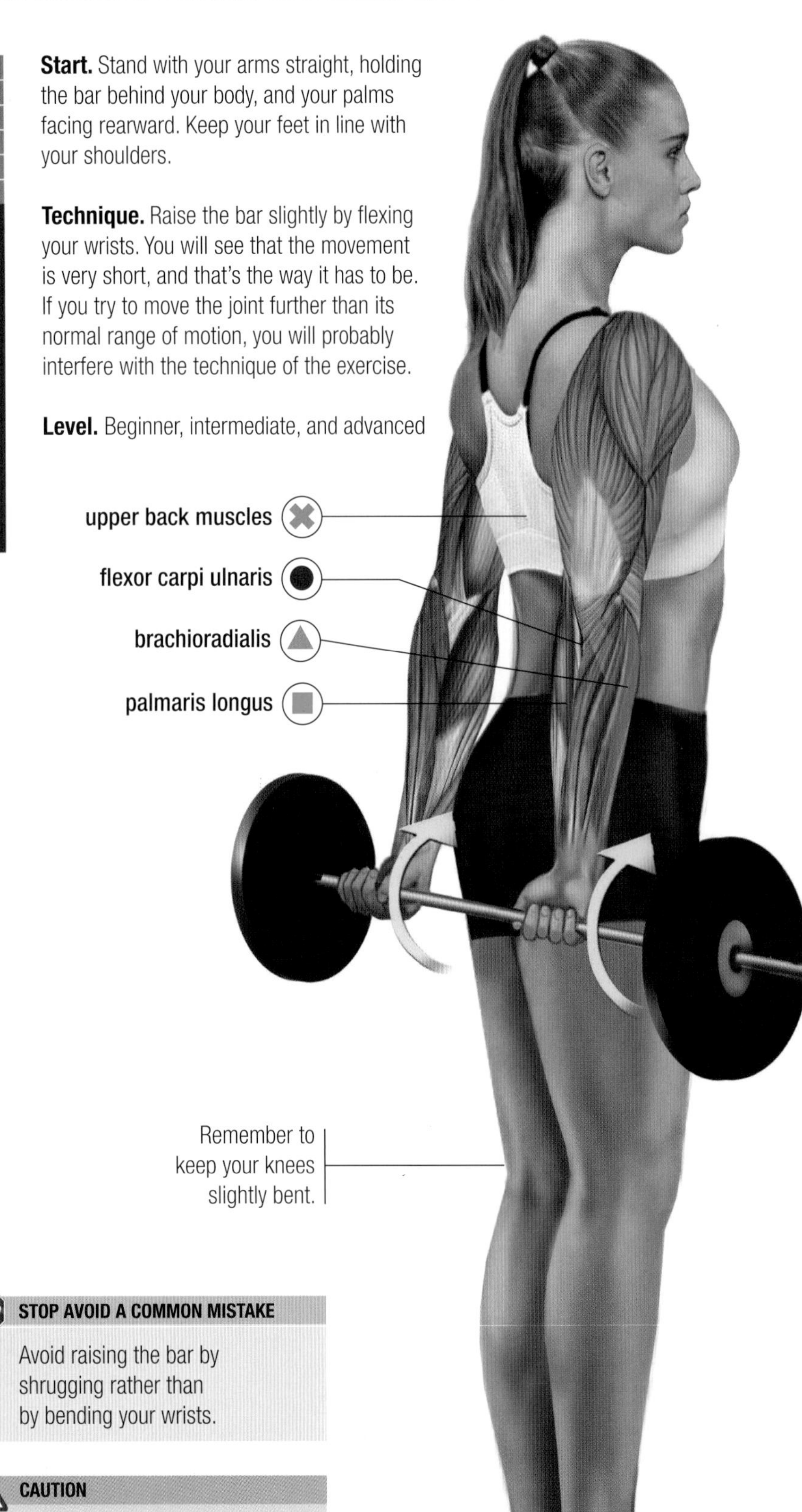

STOP AVOID A COMMON MISTAKE

Avoid raising the bar by shrugging rather than by bending your wrists.

CAUTION

When you lift the weight from the floor, keep your back straight and bend your knees. Avoid bending your back to lift the weight from the floor or to lower it when you are done.

Pronation Wrist Curls

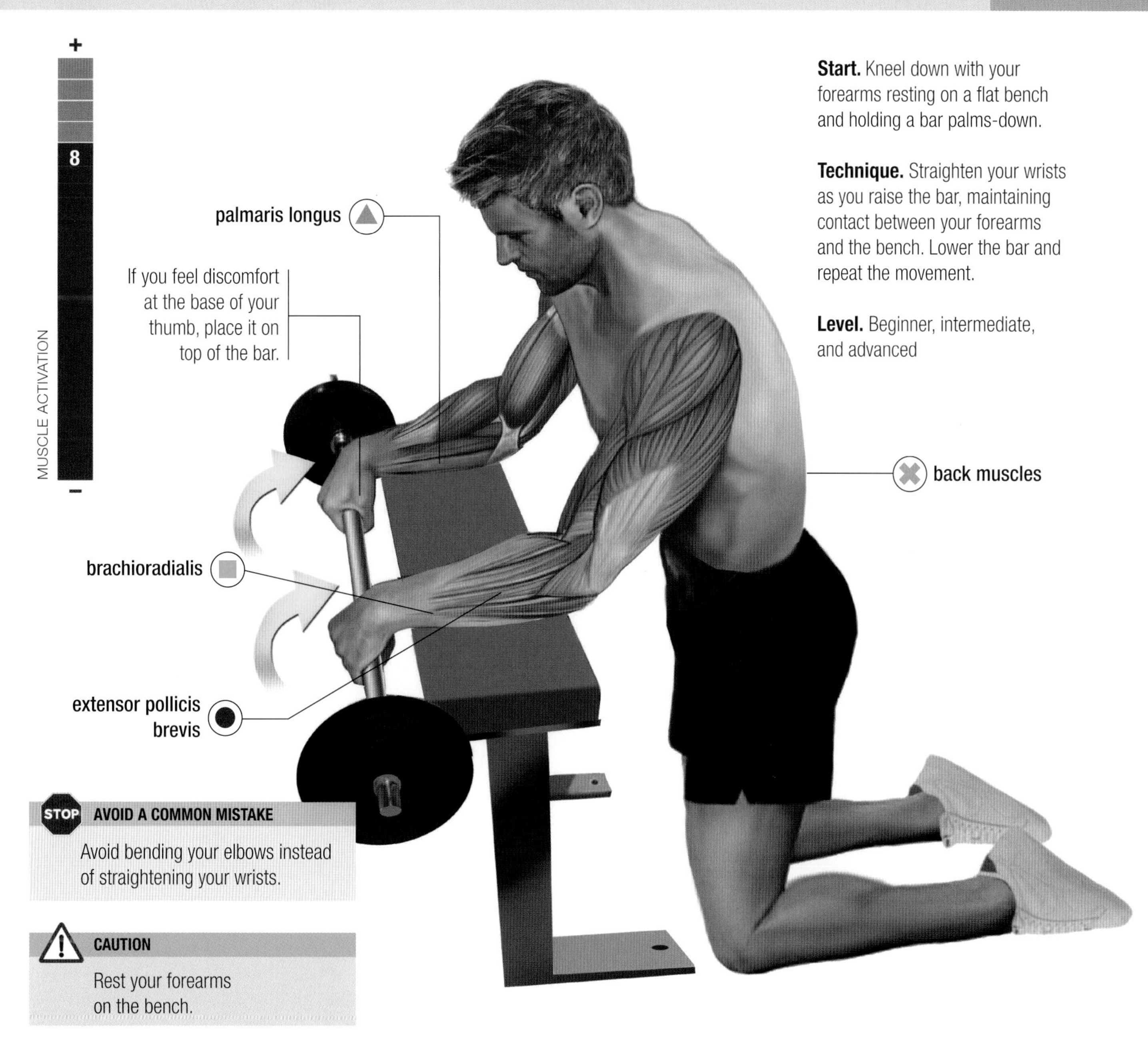

Start. Kneel down with your forearms resting on a flat bench and holding a bar palms-down.

Technique. Straighten your wrists as you raise the bar, maintaining contact between your forearms and the bench. Lower the bar and repeat the movement.

Level. Beginner, intermediate, and advanced

AVOID A COMMON MISTAKE

Avoid bending your elbows instead of straightening your wrists.

CAUTION

Rest your forearms on the bench.

VARIANT

Forearms

Supination Wrist Curls

Start. Sit with your forearms supported on your thighs, holding a bar with your palms up and your wrists extended.

Technique. Raise the bar solely by bending your wrists, and then extend your wrists again.

Level. Beginner, intermediate, and advanced

AVOID A COMMON MISTAKE

Keep your arms in a stable position.

CAUTION

Rest your forearms on your thighs to avoid placing pressure on your back.

VARIANT

Reverse-grip Barbell Curls

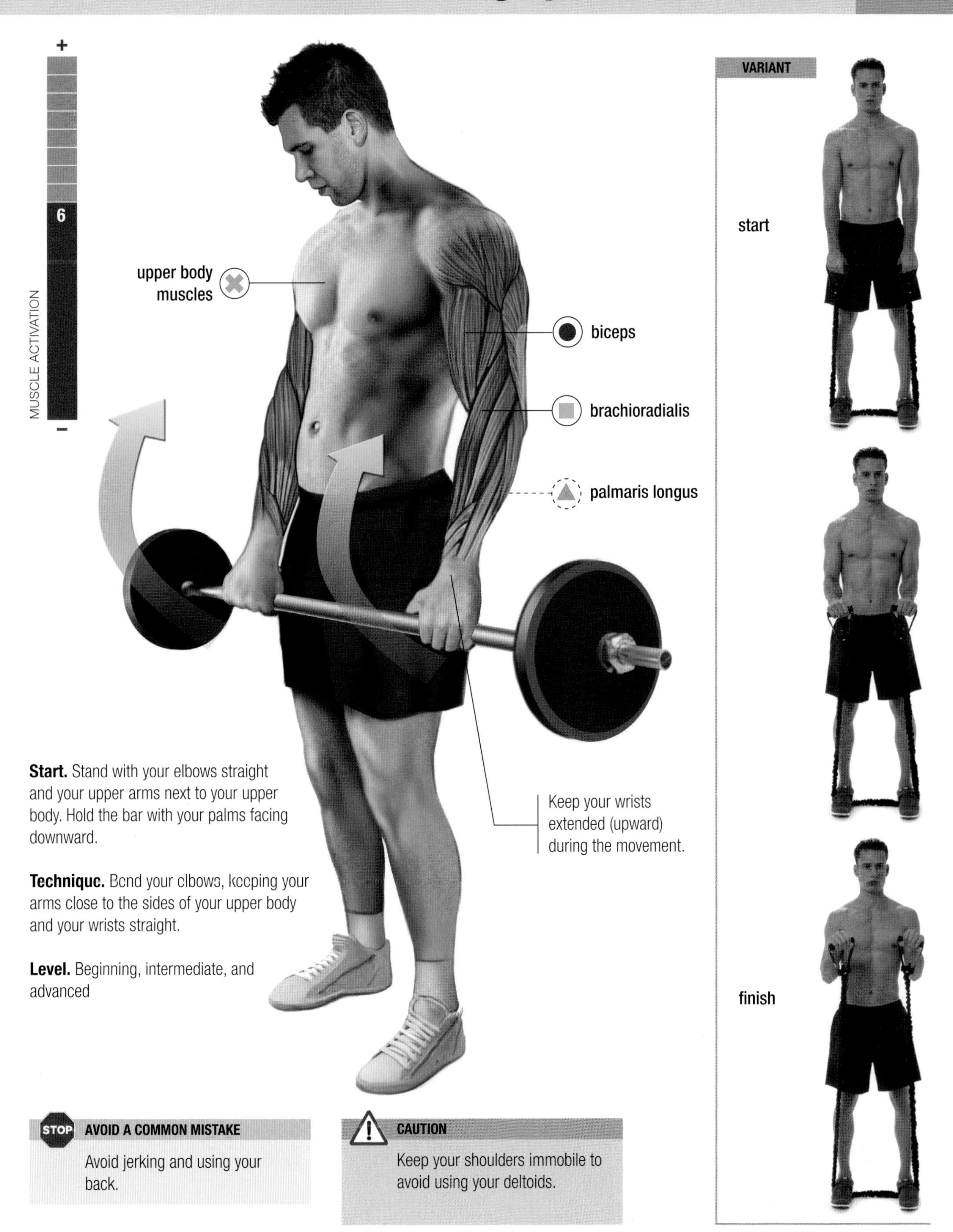

Start. Stand with your elbows straight and your upper arms next to your upper body. Hold the bar with your palms facing downward.

Technique. Bend your elbows, keeping your arms close to the sides of your upper body and your wrists straight.

Level. Beginning, intermediate, and advanced

AVOID A COMMON MISTAKE

Avoid jerking and using your back.

CAUTION

Keep your shoulders immobile to avoid using your deltoids.

Legs

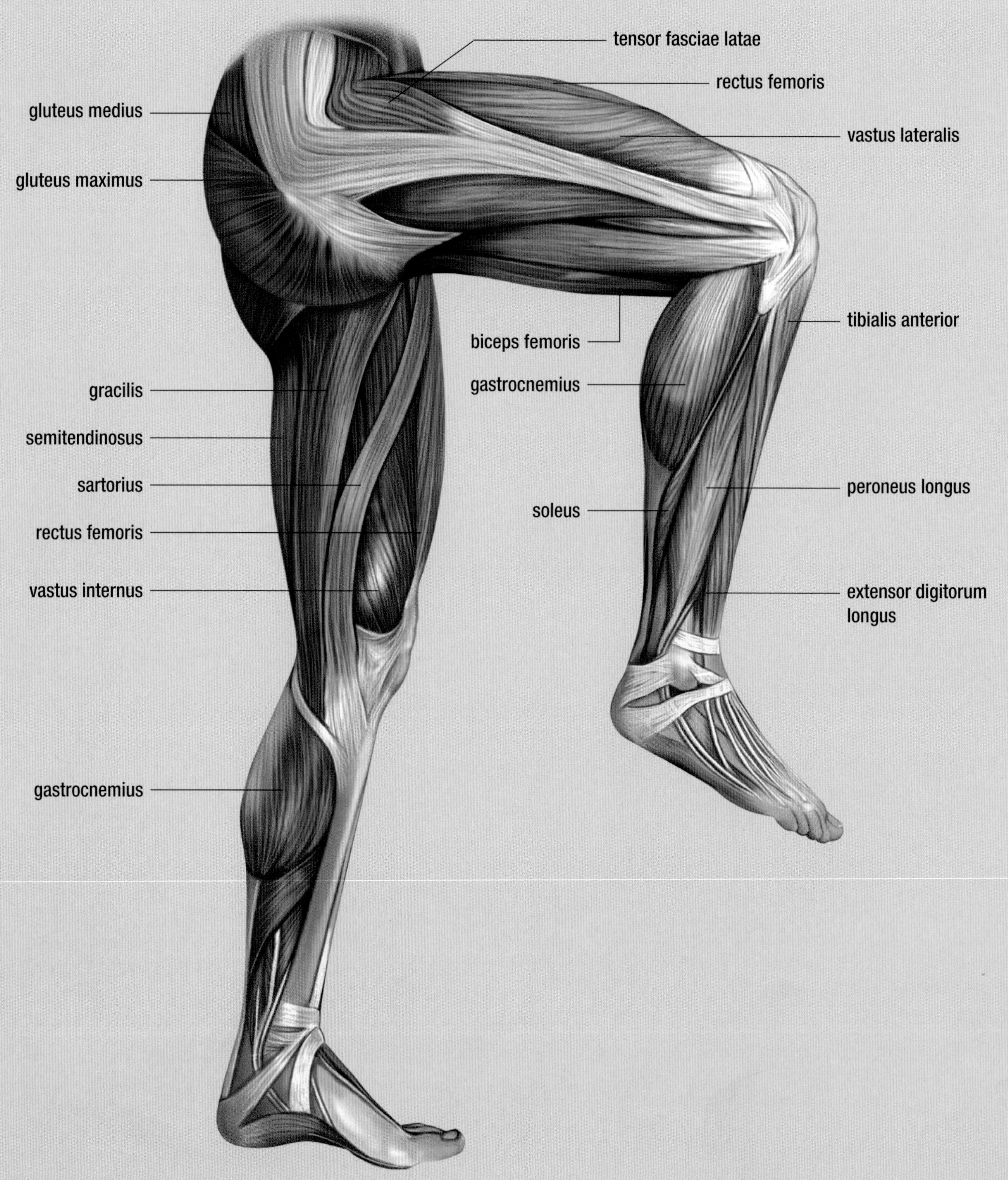
tensor fasciae latae
rectus femoris
gluteus medius
vastus lateralis
gluteus maximus
tibialis anterior
biceps femoris
gastrocnemius
gracilis
semitendinosus
sartorius
peroneus longus
soleus
rectus femoris
vastus internus
extensor digitorum longus
gastrocnemius

Technically, the leg is the part of your lower extremity between the knee and the ankle; but, this book refers to it as the area between the hip and the ankle. The legs are a part of the body that bodybuilders often neglect. However, an aesthetically pleasing, functionally balanced body needs well-developed legs. Aesthetically, they balance the size of the upper and lower parts of the body, and, on a functional level, your lower extremities are the ones that make it possible for you to move by walking, running, jumping, or bicycling, and, without going into further detail, to participate in a great variety of sports, whether individual or team.

Quadriceps. This is the largest muscle in the body; it is composed of four sections that meet at the front of the thigh.

Rectus femoris. This muscle arises at the lower anterior iliac spine. It inserts along with the other three segments at the quadricipital tendon that goes all the way to the kneecap, where it becomes known as the patellar tendon, which in turn inserts into the anterior tuberosity of the tibia.
Rectus femoris. This muscle arises on the proximal epifisis of the femur.
Vastus externus and internus. Both of these muscles originate at the epifisis and the proximal third of the femur.

The quadriceps is a very powerful muscle that enables you to remain standing and to move around, because its function is the straightening of the knee. This muscle is very important in all sports that involve moving or jumping. It is used in all athletic disciplines, but especially in the long jump, high jump, triple jump, and all running events. It is also used extensively in all team sports, especially in football, soccer, volleyball, and basketball.

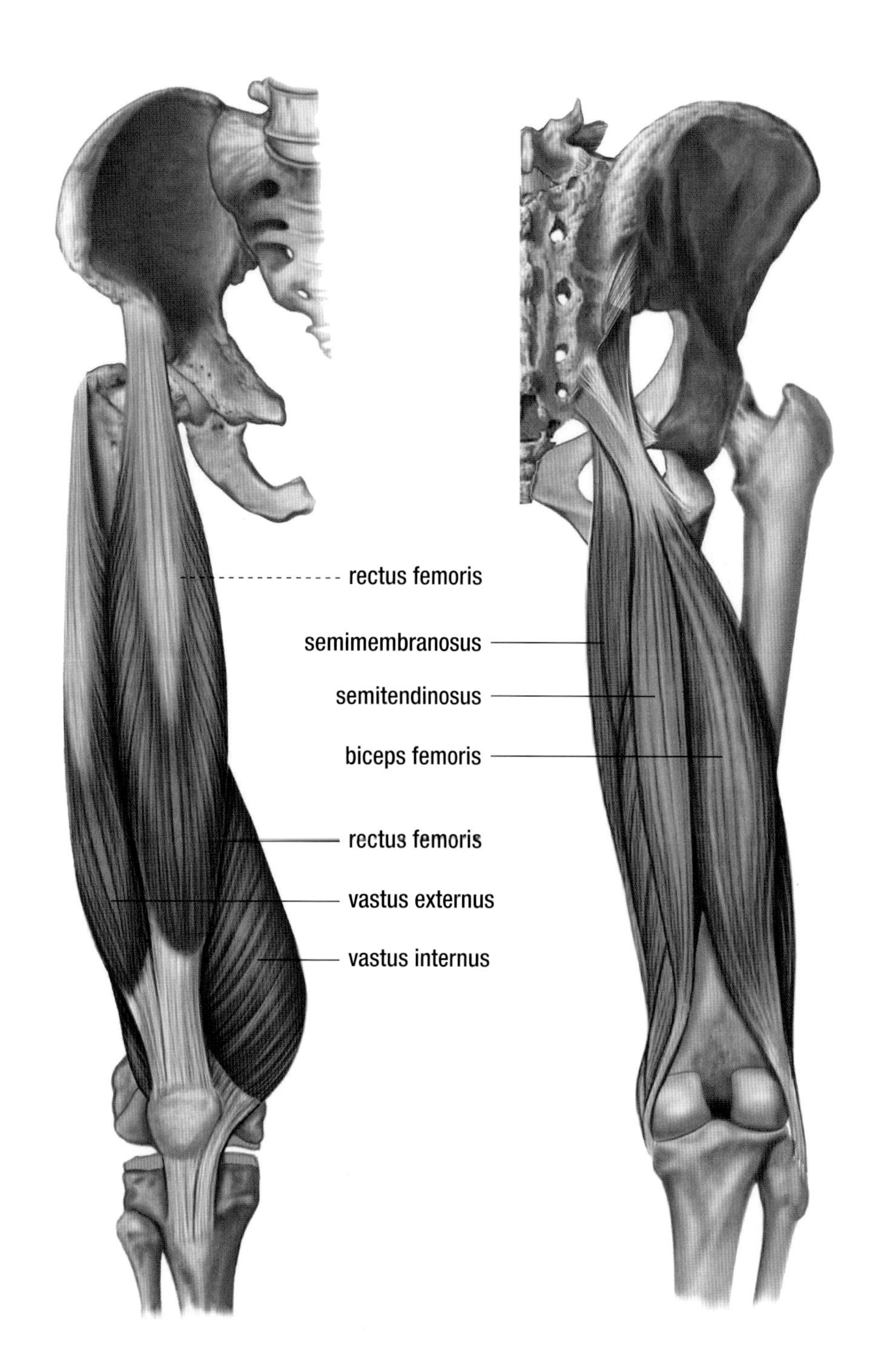

Ischiotibial muscles. The ischiotibial muscles are three muscles that are found in the rear of the thigh. Their function is to bend the knee.

Biceps femoris. This muscle arises at the ischium and the femoral diaphysis, and it inserts at the proximal epiphysis of the tibia and fibula.
Semitendinosus. This muscle arises at the ischium and inserts at the proximal part of the tibial diaphysis.
Semimembranosus. This muscle arises at the ischium and inserts at the proximal epiphysis of the tibia.

These muscles are used in rapid movements, such as the quick acceleration at the start of a race, so it is common for them to experience injury in fast-moving sports, with sudden starts and stops and sudden changes of direction, such as soccer, tennis, and football.

Gastrocnemius. Commonly referred to as the "calf," this muscle is located in the lower rear part of the leg. It has a medial and a lateral part, and it arises in the sides of the distal epiphysis of the femur. It inserts with the Achilles tendon, which it shares with the soleus, in the rear portion of the heel bone. Its function is plantar bending at the ankle. It allows you to stand on tiptoes, so it is essential in disciplines such as rhythmic and athletic gymnastics and ballet.

Soleus. This muscle is located in the back of the leg. Its main function is the plantar bending of the ankle, like the gastrocnemius. In fact, the set of the two muscles is often referred to as the sural triceps. The soleus arises in the proximal epiphysis of the fibula and in the diaphysis of the tibia and fibula. Its insertion is shared with the gastrocnemius, and it is used in the same activities.

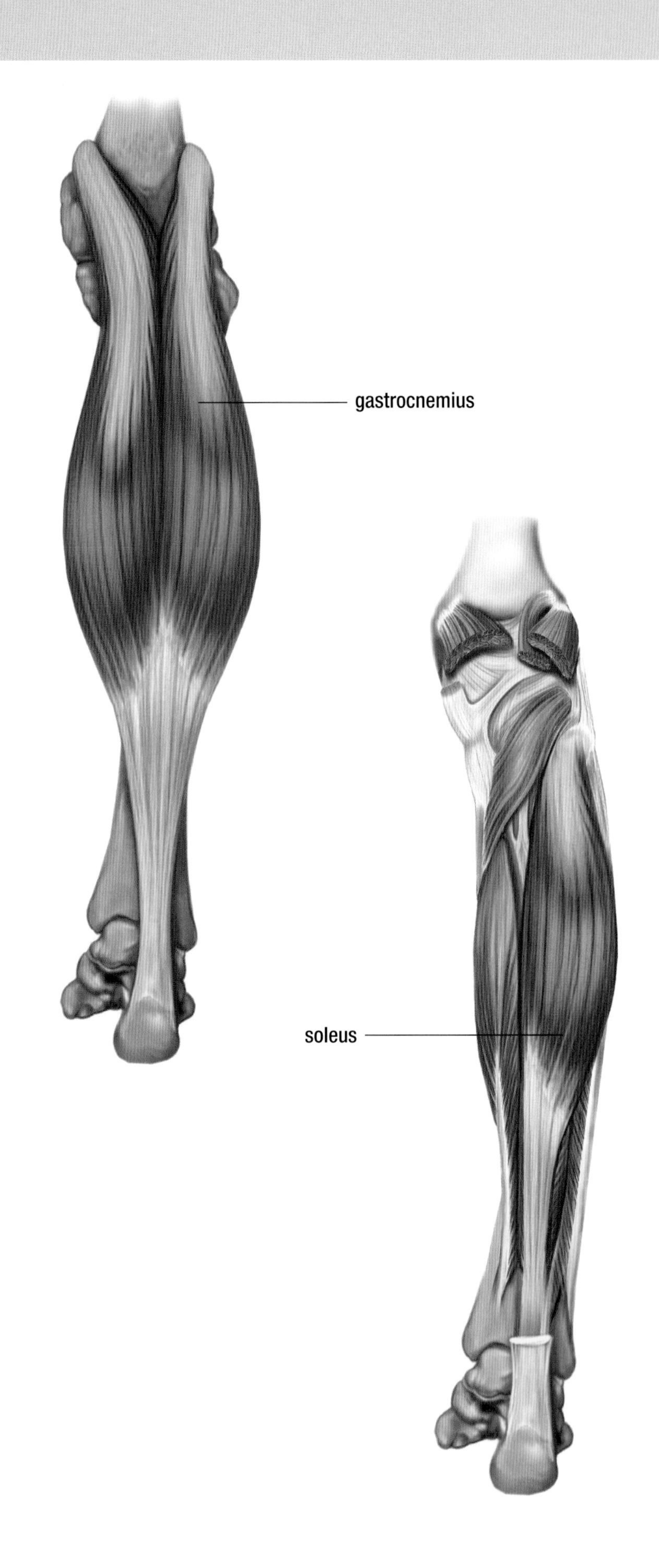

Abductors. Even though there are muscles that have the same function, when this book speaks of abductors of the hip, it refers mainly to the tensor fasciae latae, which arises at the iliac crest, fascia lata, and iliac spine. Its insertion is in the proximal epiphysis of the tibia. Its function is abduction of the hip—in other words, to move the lower extremity outward from the median line of the body. It allows you to separate your legs, and it is used in the lateral movements of judo and in the movements and kicks of tae kwan do and karate.

Adductors. These are the adductor magnus, the adductor brevis, and the adductor longus. They originate in the pubis, and they insert along the femoral diaphysis. Their function is adduction of the hip—in other words, to move the lower extremity toward the midline of the body. They allow you to close your legs, so they are particularly important in sports such as judo (in the techniques that involve the legs) and soccer (passing on the inside).

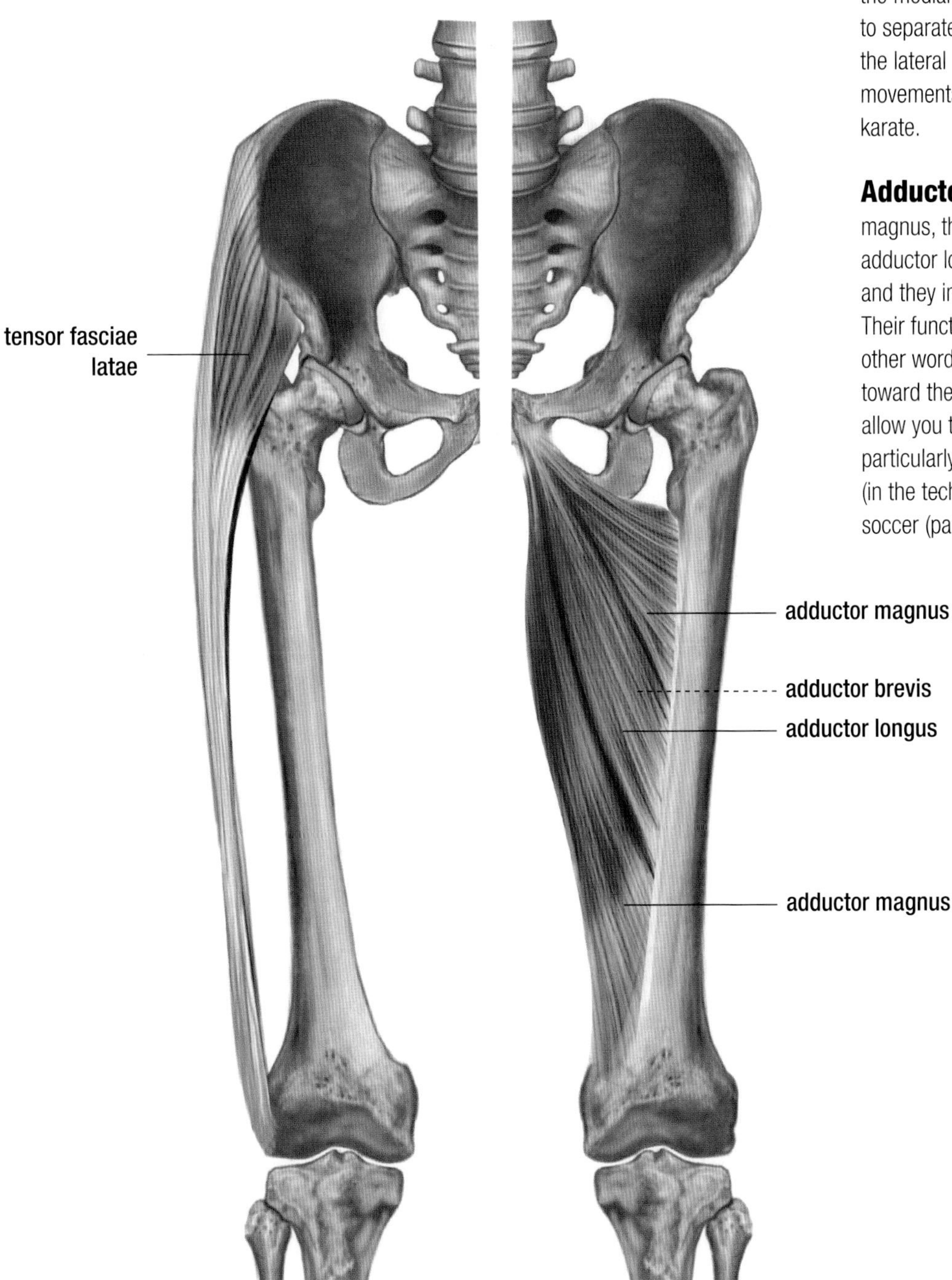

Quadriceps

Squats

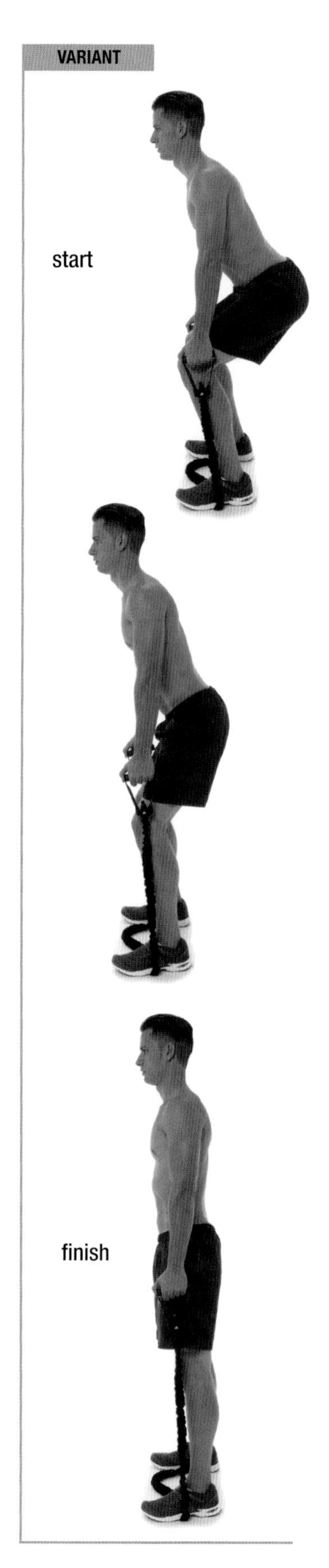

Start. Stand with your feet shoulder-width apart and your knees straight but not locked. Place the bar onto your shoulders, behind your neck, and hold it with both hands.

Technique. Bend your knees to 90° as you lean your upper body forward, keeping your back straight. Descend slowly and return to the starting position.

Level. Advanced

AVOID A COMMON MISTAKE

Avoid bending the knees beyond 90°. Otherwise, you could end up damaging your menisci.

CAUTION

You can lean your upper body forward slightly at the lowest stage of the movement to maintain your balance, but avoid arching your back.

Leg Press

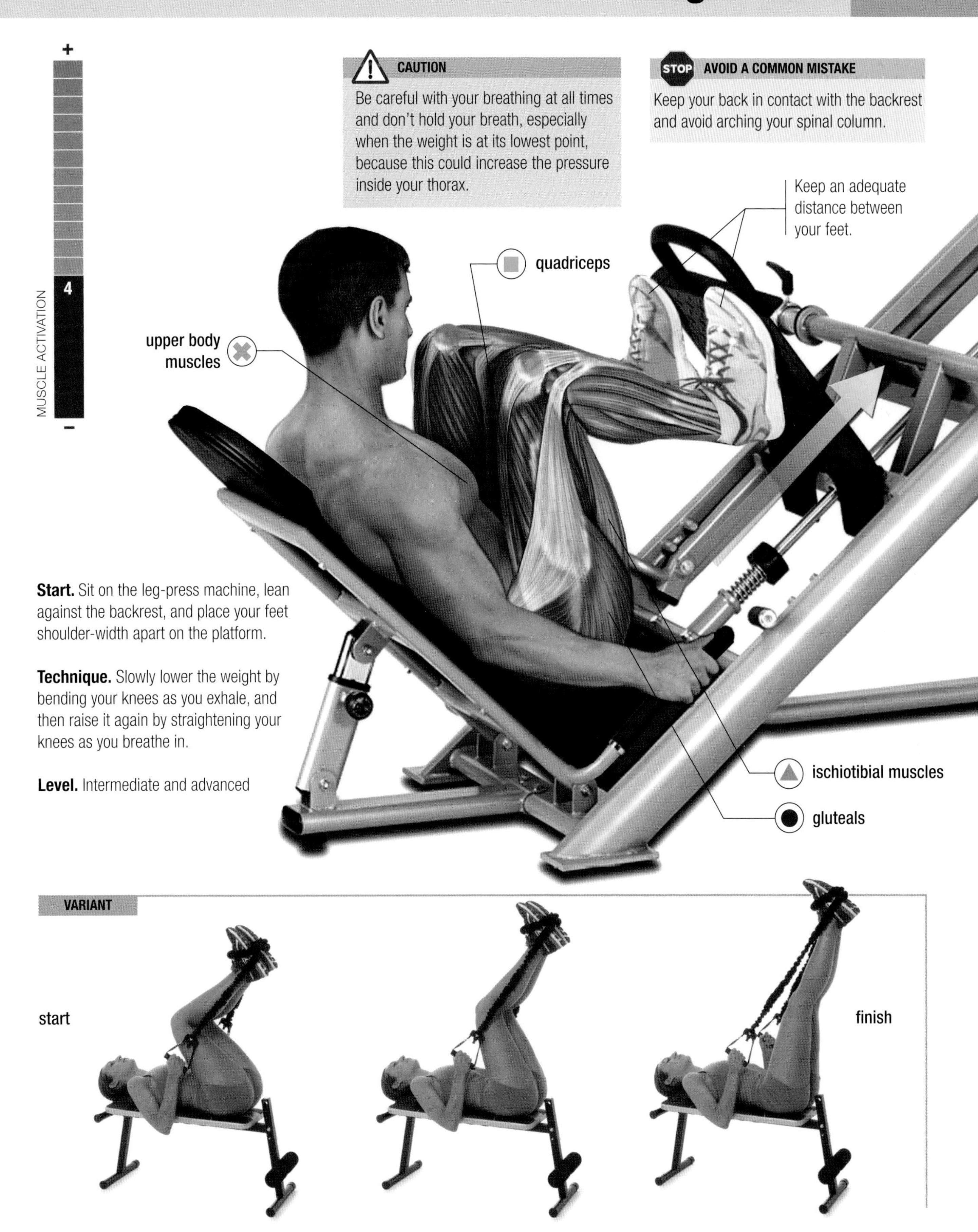

+
4
MUSCLE ACTIVATION
–
CAUTION
Be careful with your breathing at all times and don't hold your breath, especially when the weight is at its lowest point, because this could increase the pressure inside your thorax.
STOP
AVOID A COMMON MISTAKE
Keep your back in contact with the backrest and avoid arching your spinal column.
Keep an adequate distance between your feet.
quadriceps
upper body muscles
ischiotibial muscles
gluteals
VARIANT
start
finish

Start. Sit on the leg-press machine, lean against the backrest, and place your feet shoulder-width apart on the platform.

Technique. Slowly lower the weight by bending your knees as you exhale, and then raise it again by straightening your knees as you breathe in.

Level. Intermediate and advanced

Quadriceps

Hack Squats

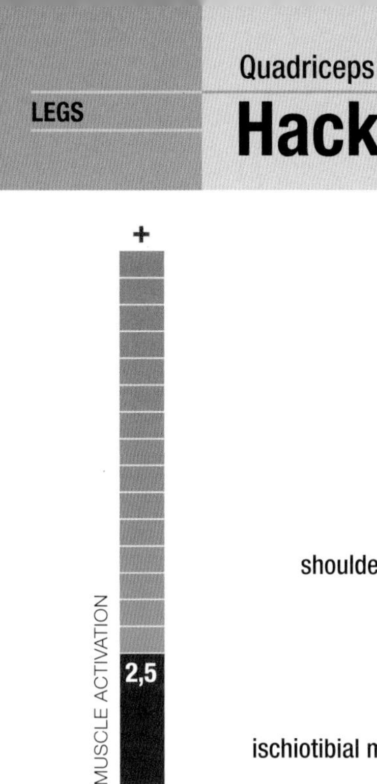

AVOID A COMMON MISTAKE

Avoid arching your back, because this could cause damage to your spine.

CAUTION

Do not bend your knees much more than 90°.

VARIANT

Start. Stand with the cushions pressing against your shoulders and your knees straight but not locked. Place your hands on the grips that usually are located in front of your shoulders or near your hips.

Technique. Lower the weight by bending your knees no lower than 90º and return to the starting point. Make sure you do not lock your knees when you reach the point of maximum extension.

Level. Intermediate and advanced

Leg Extension

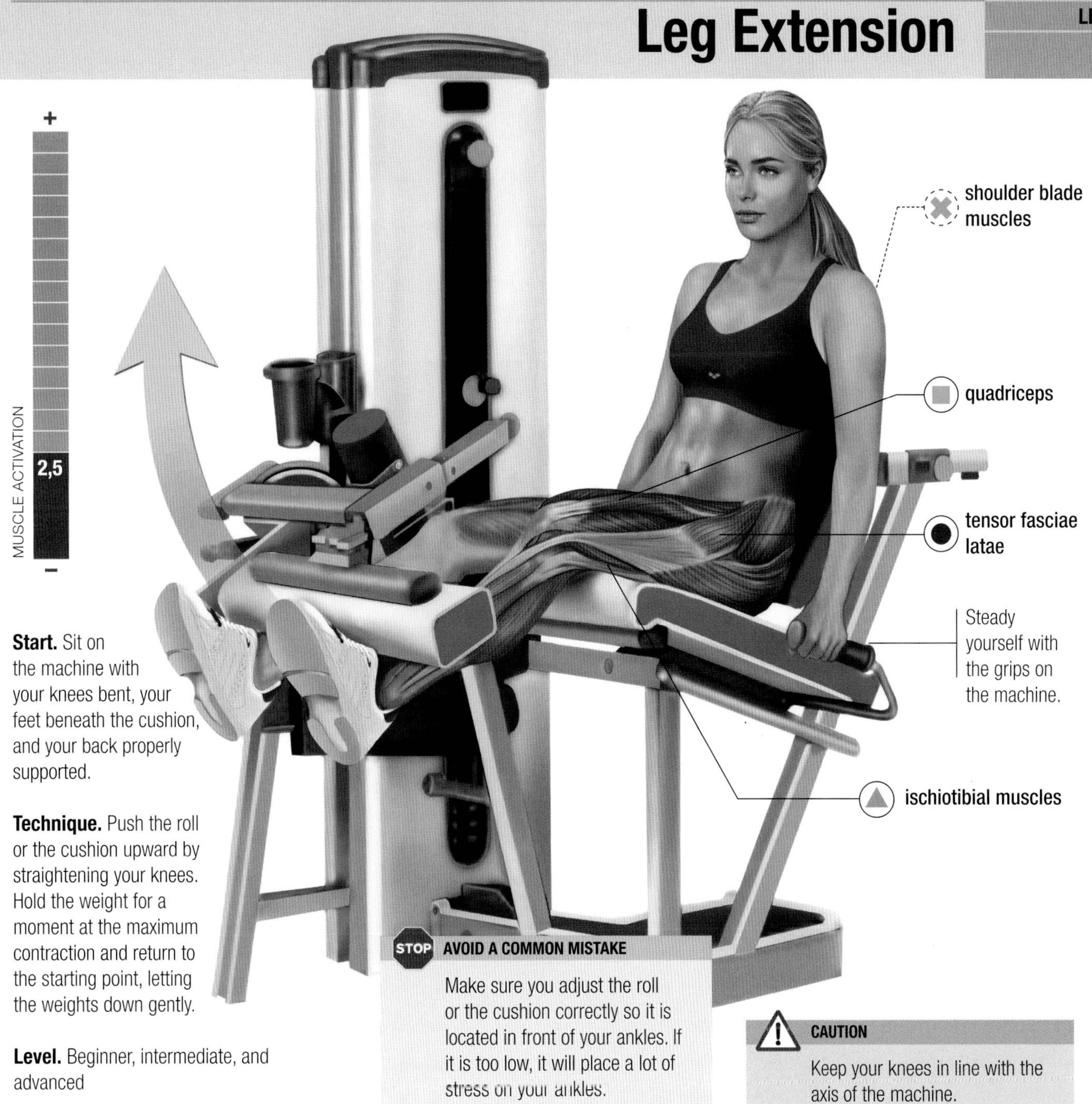

Start. Sit on the machine with your knees bent, your feet beneath the cushion, and your back properly supported.

Technique. Push the roll or the cushion upward by straightening your knees. Hold the weight for a moment at the maximum contraction and return to the starting point, letting the weights down gently.

Level. Beginner, intermediate, and advanced

STOP AVOID A COMMON MISTAKE

Make sure you adjust the roll or the cushion correctly so it is located in front of your ankles. If it is too low, it will place a lot of stress on your ankles.

CAUTION

Keep your knees in line with the axis of the machine.

VARIANT

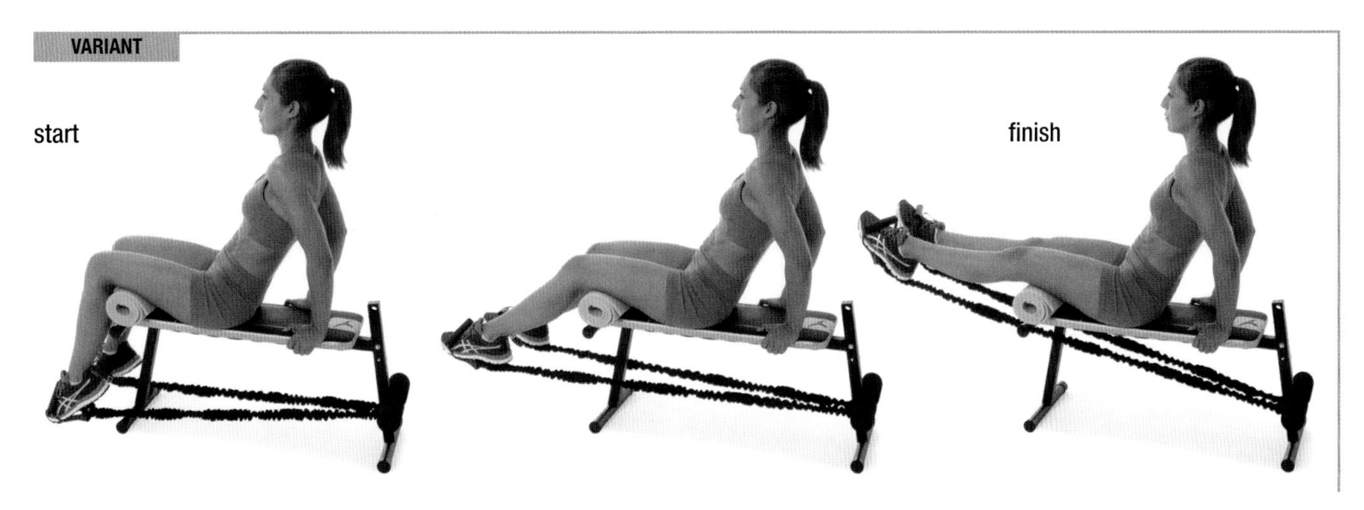

Quadriceps

Canadian Squats

shoulder blade muscles
You can use a cushion or a towel to relieve the pressure from the bar on your shoulders, especially if you are working with heavy weights.
quadriceps
+
2,5
MUSCLE ACTIVATION
gluteals
ischiotibial muscles

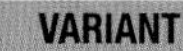
VARIANT

start

finish

Start. Stand with your feet in line with your shoulders. Rest a bar on the front of your shoulders, holding your elbows to the front for balance, as shown in the illustration.

Technique. Lower your body by bending at the hips and knees until both are bent about 90°. Keep the movement slow and controlled, and return to the starting point, without locking your knees.

Level. Advanced

AVOID A COMMON MISTAKE

Avoid bending your knees too far to squat deeper, because this can damage your menisci and put too much tension on your ankles.

CAUTION

Keep your back straight throughout the exercise so that you maintain the natural curvature of your spine.

Hamstring Curls

Start. Lie face-down on the bench. Keep your legs straight and the cushion in the area behind your ankles.

Technique. Lift the cushion by bending your knees. Once you have lifted the weight, slowly reverse the movement without letting the weights clank together.

Level. Beginner, intermediate, and advanced

CAUTION

Be careful with the speeds and the weights, because these muscles are not especially powerful; therefore, they are more susceptible to injury with sudden movements.

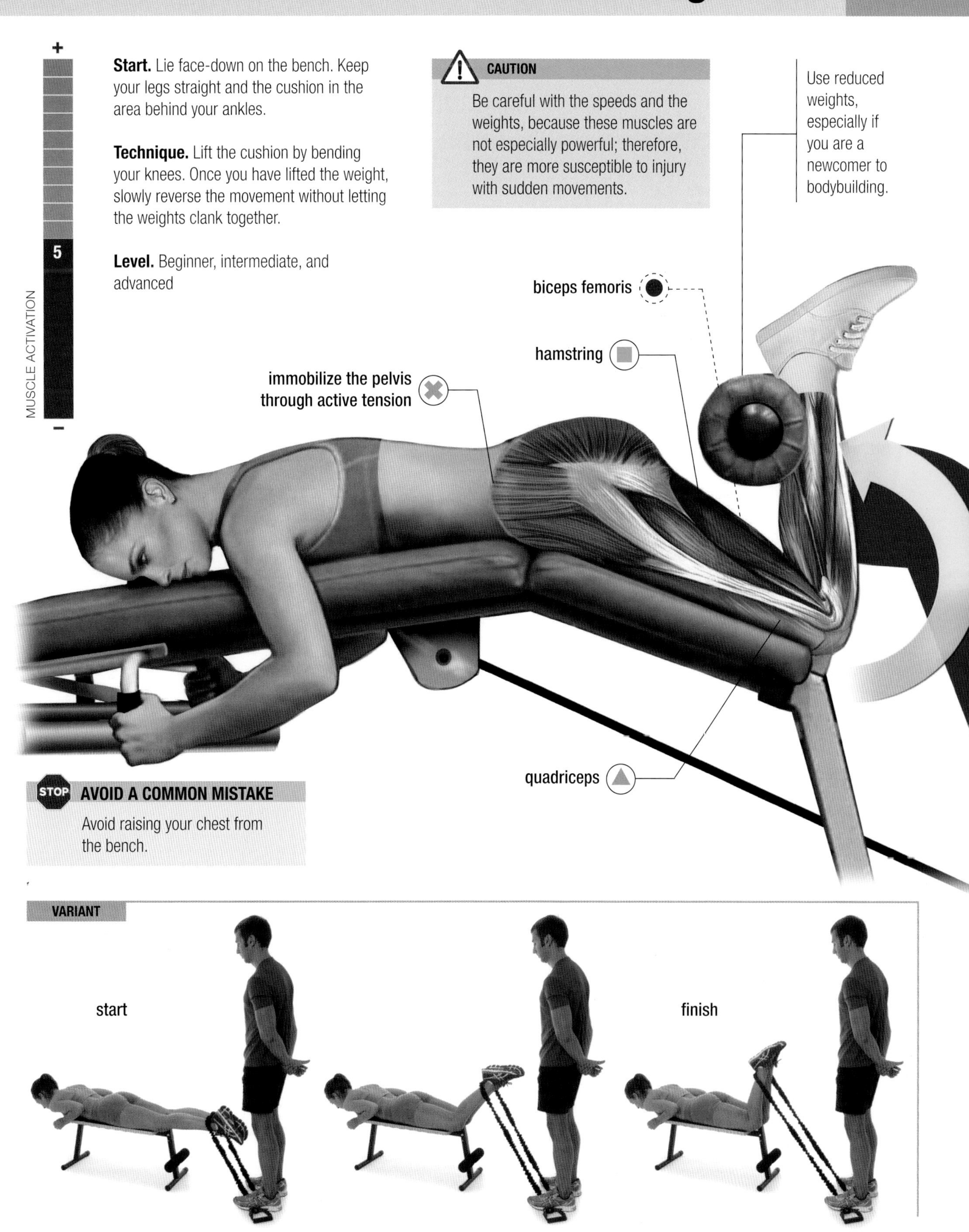

AVOID A COMMON MISTAKE

Avoid raising your chest from the bench.

VARIANT

Hamstrings

Seated Hamstring Curls

Start. Sit with your knees straight and the stationary cushion higher up. The movable roll cushion should be behind your ankles.

AVOID A COMMON MISTAKE

Avoid placing the stationary cushion right on top of your knees; it must press down higher up than your knees, at the lower part of your thigh.

CAUTION

Be careful with the weights at the outset, until you learn how to interpret the sensations you feel while doing the exercise.

Technique. Bend your knees to 90°, hold the position, and slowly return to the starting point without jerking.

Level. Beginner, intermediate, and advanced

VARIANT

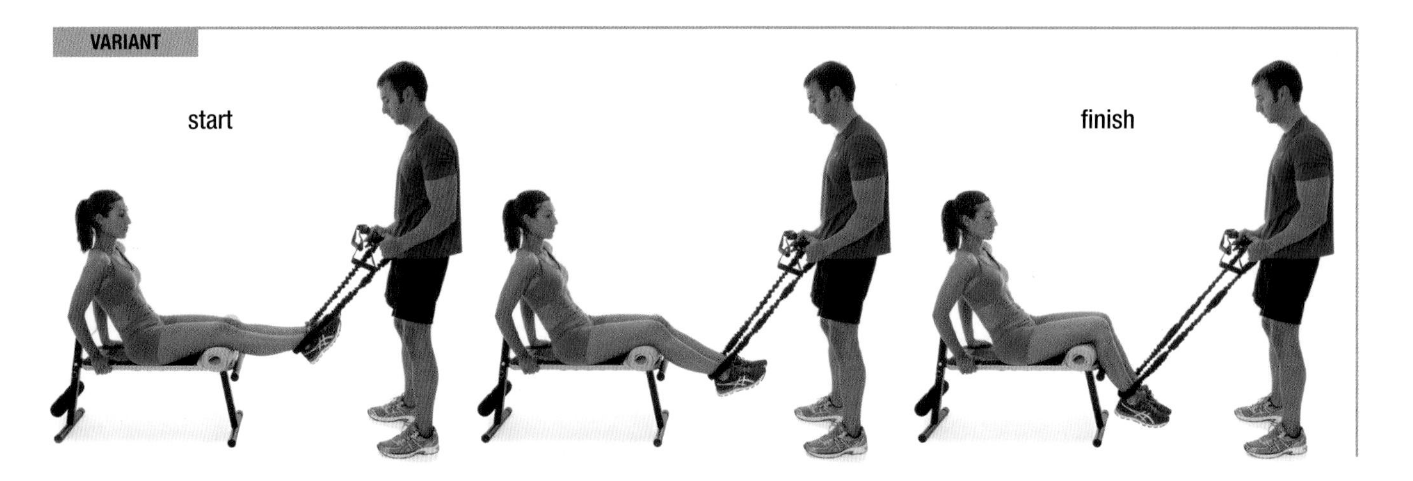

Barbell Toe Raises

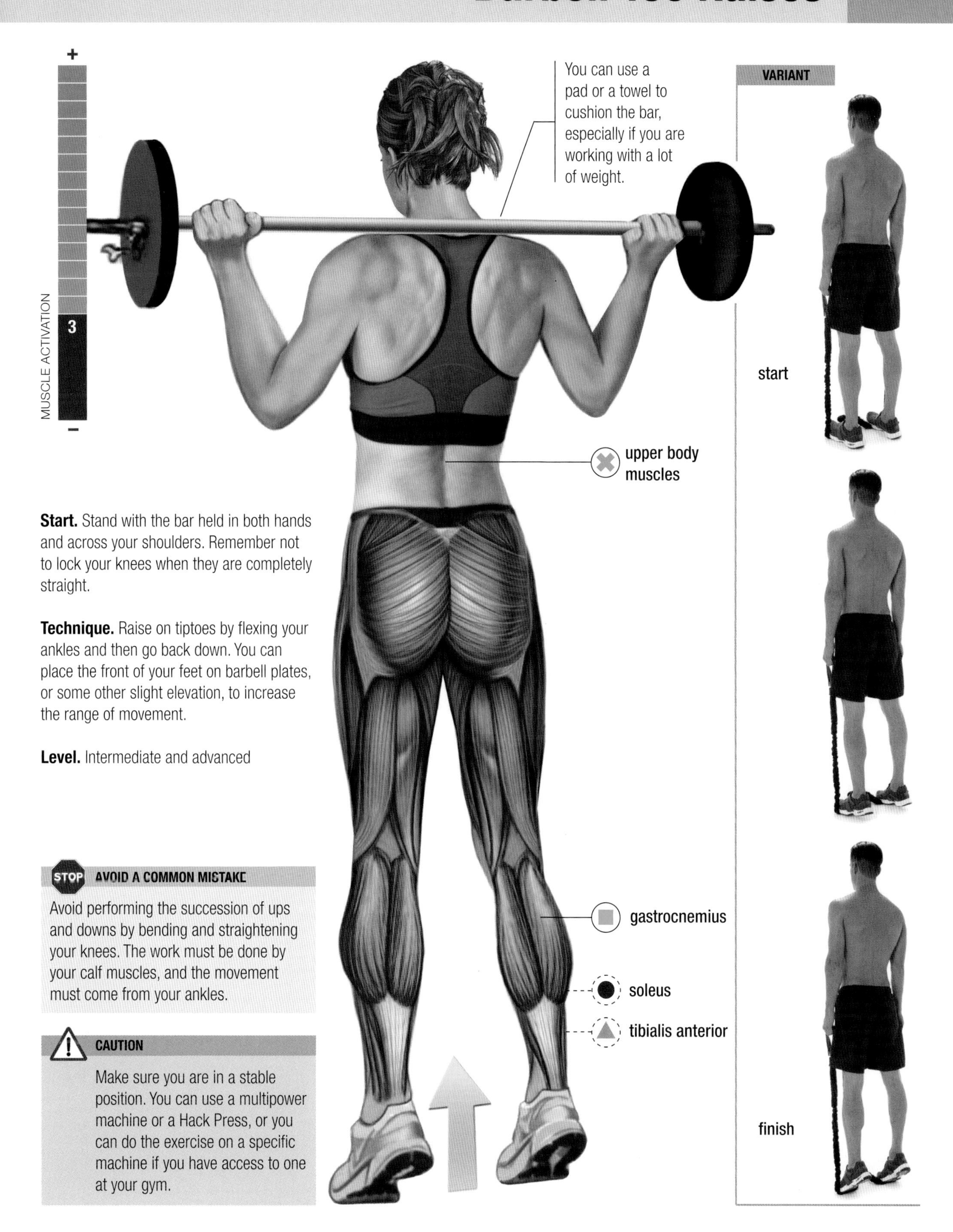

+
MUSCLE ACTIVATION
3
–
You can use a pad or a towel to cushion the bar, especially if you are working with a lot of weight.
VARIANT
start
finish
upper body muscles
gastrocnemius
soleus
tibialis anterior

Start. Stand with the bar held in both hands and across your shoulders. Remember not to lock your knees when they are completely straight.

Technique. Raise on tiptoes by flexing your ankles and then go back down. You can place the front of your feet on barbell plates, or some other slight elevation, to increase the range of movement.

Level. Intermediate and advanced

STOP AVOID A COMMON MISTAKE

Avoid performing the succession of ups and downs by bending and straightening your knees. The work must be done by your calf muscles, and the movement must come from your ankles.

CAUTION

Make sure you are in a stable position. You can use a multipower machine or a Hack Press, or you can do the exercise on a specific machine if you have access to one at your gym.

Gastrocnemius

Barbell Toe Raises on One Foot

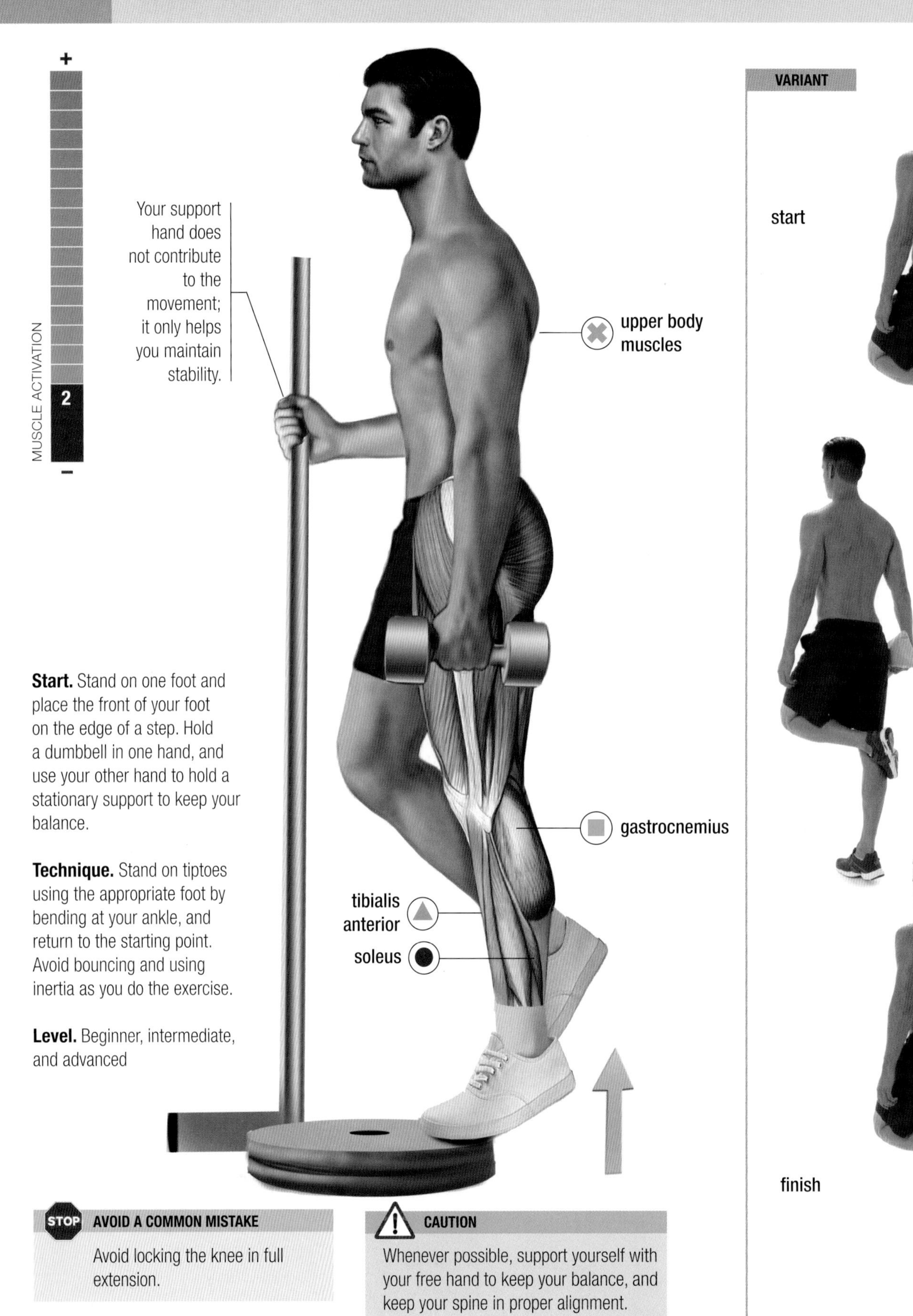

Start. Stand on one foot and place the front of your foot on the edge of a step. Hold a dumbbell in one hand, and use your other hand to hold a stationary support to keep your balance.

Technique. Stand on tiptoes using the appropriate foot by bending at your ankle, and return to the starting point. Avoid bouncing and using inertia as you do the exercise.

Level. Beginner, intermediate, and advanced

AVOID A COMMON MISTAKE

Avoid locking the knee in full extension.

CAUTION

Whenever possible, support yourself with your free hand to keep your balance, and keep your spine in proper alignment.

Soleus on the Machine

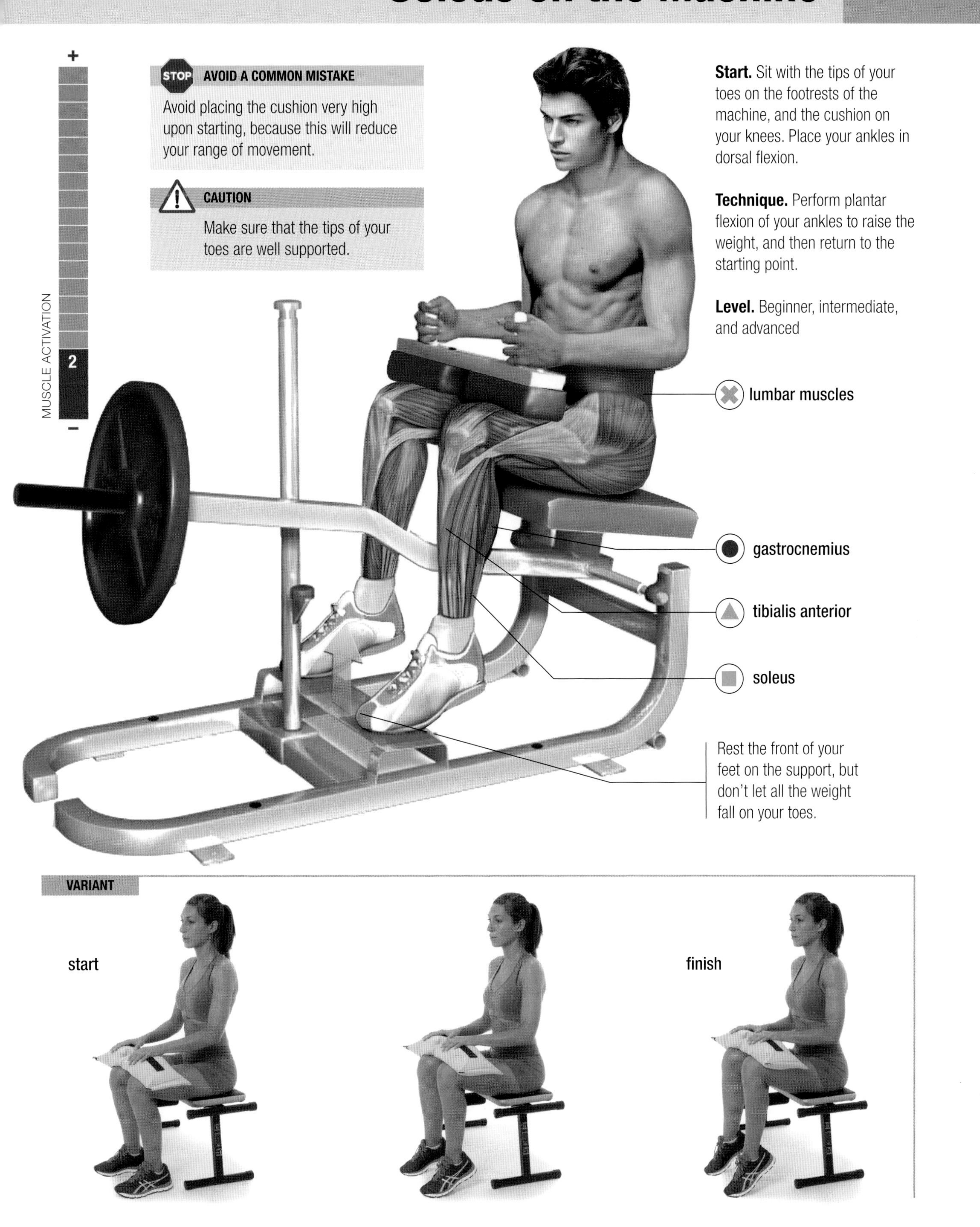
+
MUSCLE ACTIVATION
2
–
lumbar muscles
gastrocnemius
tibialis anterior
soleus
Rest the front of your feet on the support, but don't let all the weight fall on your toes.
VARIANT
start
finish

STOP AVOID A COMMON MISTAKE

Avoid placing the cushion very high upon starting, because this will reduce your range of movement.

CAUTION

Make sure that the tips of your toes are well supported.

Start. Sit with the tips of your toes on the footrests of the machine, and the cushion on your knees. Place your ankles in dorsal flexion.

Technique. Perform plantar flexion of your ankles to raise the weight, and then return to the starting point.

Level. Beginner, intermediate, and advanced

Abductors

Abductors on the Machine

+
STOP
AVOID A COMMON MISTAKE
Avoid pushing your hips forward, and press tightly against the backrest.
Do the exercise slowly and deliberately, and work with light or moderate weights.
adductors
MUSCLE ACTIVATION
3
−
lumbar muscles
abductors
gluteus medius and minimus

CAUTION

Place the outside of your knees at the right spot on the cushions, using the greatest support surface, and keep your feet stationary on the footrests.

Start. Sit with your legs together and your knees bent. The cushions should touch the outer part of your knees, and your feet will be on a movable support.

Technique. Move your legs apart by means of hip abduction. Once you reach the point of maximum separation, hold the position, and then return to the start as you control the descent of the weights.

Level. Beginner, intermediate, and advanced

VARIANT

start
finish

Adductors on the Machine

Start. Sit with your legs apart and the insides of your knees pressed against the cushions. Keep your feet stationary on the footrests.

Technique. Bring your legs together by means of hip adduction until the cushions are together. Then return to the starting point, controlling the speed of the exercise so that you can stop it immediately at any point in its range.

Level. Beginner, intermediate, and advanced

lumbar muscles

adductors

pectinius

abductors

Adjust the opening angle of the machine so that it causes you no trouble or discomfort.

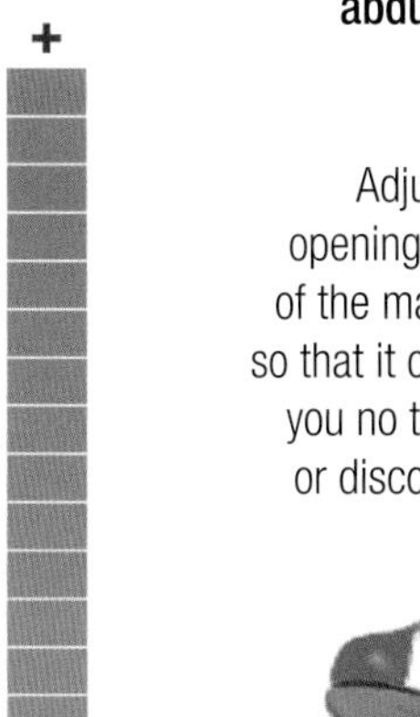

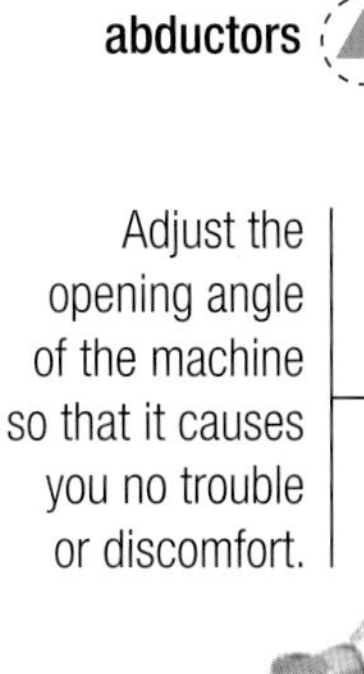

+

3

MUSCLE ACTIVATION

–

STOP AVOID A COMMON MISTAKE

Avoid short movements. It is better to reduce the weight and do more extensive movements.

CAUTION

Adjust the weight and the maximum spread on the machine. Avoid injuries.

VARIANT

Gluteals

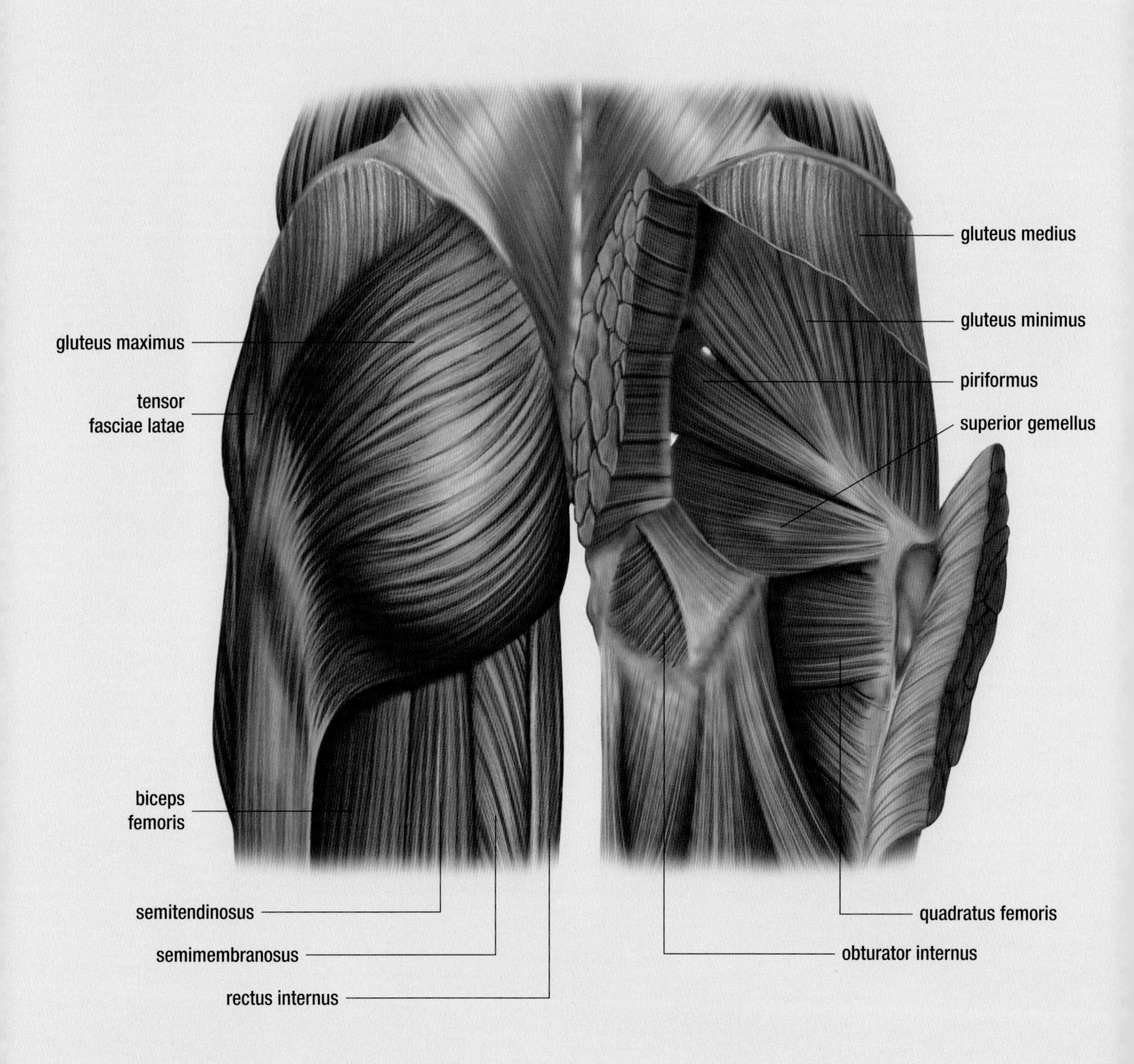
gluteus medius
gluteus minimus
gluteus maximus
piriformus
tensor
fasciae latae
superior gemellus
biceps
femoris
semitendinosus
quadratus femoris
semimembranosus
obturator internus
rectus internus

These muscles are found in the region of the buttocks. They are the gluteus maximus, medius, and minimus. This section of the book will show you primarily how to work the gluteus maximus, because it is the largest of the three and the main muscle in straightening the hip. The gluteus medius and minimus participate in hip abduction, so this book works both of them into the exercises in this section and in the abductor exercises.

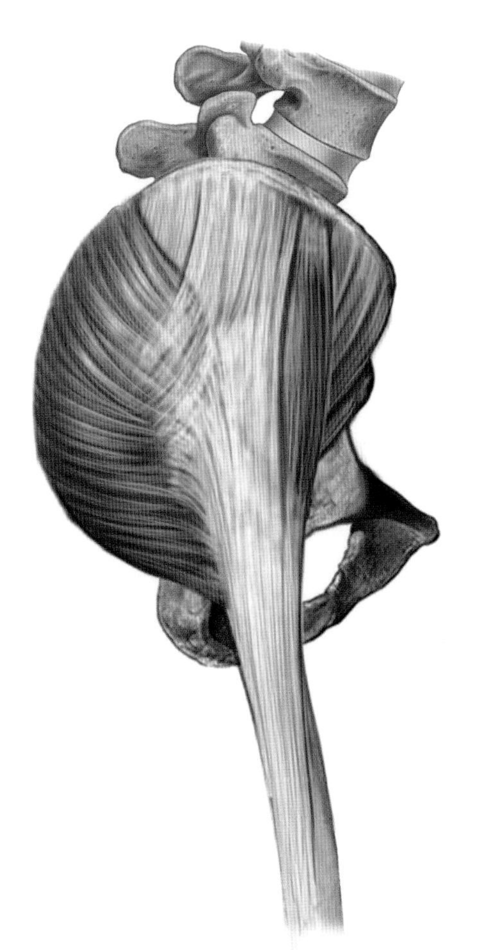

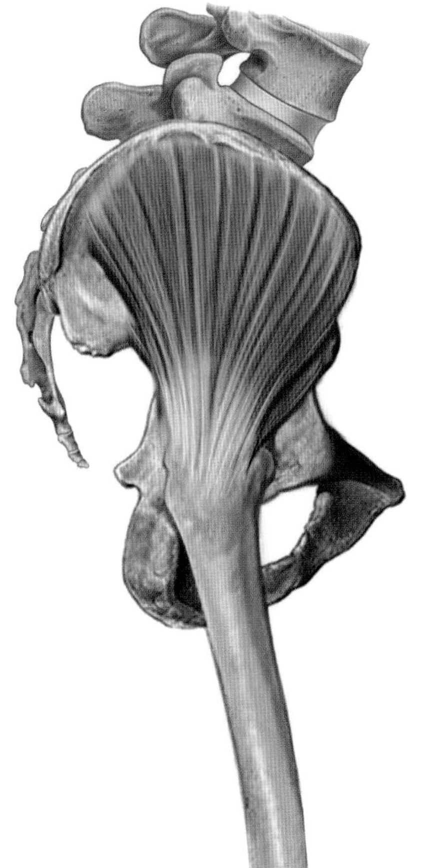

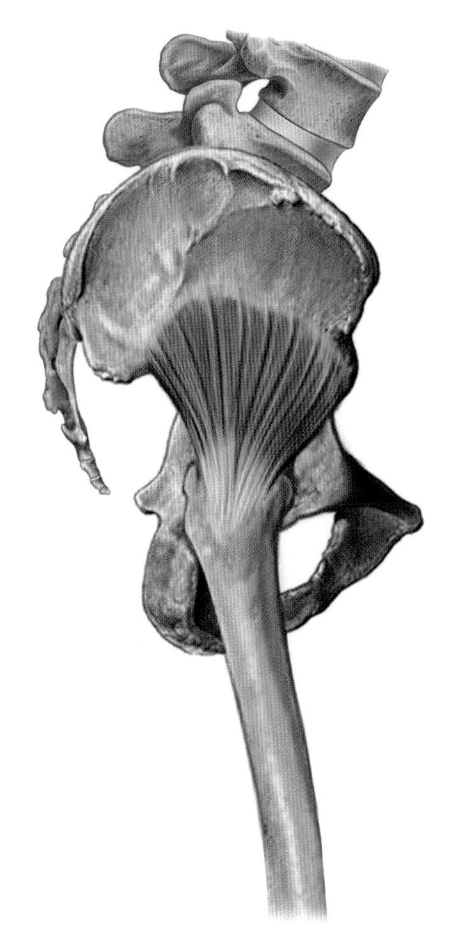

Gluteus maximus. This muscle arises at the ileum, sacrum, and coccyx bones, and it inserts at the proximal third of the femur. Its main function is straightening the hip, and, as a result, it is used when you perform a jump or a fast start. This means that the gluteus maximus is a very important muscle in many athletic performances, such as the 100- and 200-meter sprints, high jump, long jump, and triple jump. It is also used in team and individual sports that involve high-speed movement, such as rugby, football, tennis, and hockey.

Gluteus medius. This muscle arises at the rear part of the iliac crest, and it inserts on the trochanter major of the femur. Its main function is abduction of the hip, so it is used in activities that require lateral movement, such as rhythmic gymnastics and dance, and in most team sports.

Gluteus minimus. This muscle arises at the external ileum, and it inserts at the trochanter major of the femur. Its main function, like the gluteus medius, is abduction of the hip, and it is used actively in the same types of activities.

Gluteus maximus

Back Kick with Pulley or Gluteal Pull with Cable

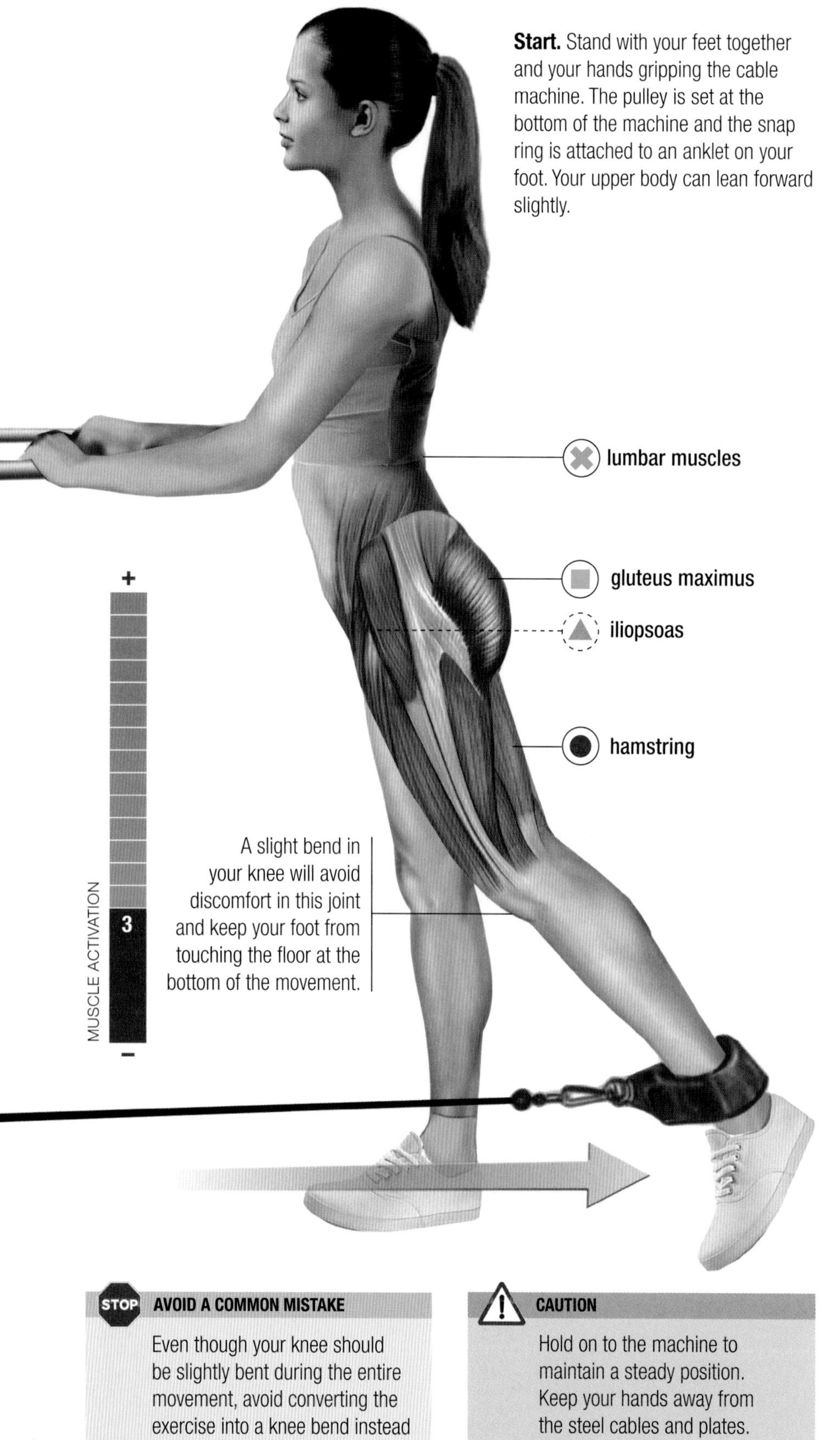
lumbar muscles
gluteus maximus
iliopsoas
hamstring
A slight bend in your knee will avoid discomfort in this joint and keep your foot from touching the floor at the bottom of the movement.
+
3
MUSCLE ACTIVATION
–

Start. Stand with your feet together and your hands gripping the cable machine. The pulley is set at the bottom of the machine and the snap ring is attached to an anklet on your foot. Your upper body can lean forward slightly.

Technique. Slowly raise your foot toward the rear by extending your hip, and return to the starting point. Remember that the knee should remain slightly bent during the entire movement. The movement involves moving your foot with the ankle rearward, like a kick.

Level. Beginner, intermediate, and advanced

VARIANT

start

finish

AVOID A COMMON MISTAKE

Even though your knee should be slightly bent during the entire movement, avoid converting the exercise into a knee bend instead of a hip extension.

CAUTION

Hold on to the machine to maintain a steady position. Keep your hands away from the steel cables and plates.

Gluteal Raise on All Fours with Ankle Weight

Start. Get down on all fours on a padded mat, with an ankle weight on one of your ankles.

Technique. Lift your foot so that the sole points upward. Then lower the foot until your leg is nearly in line with the other one, but without touching the floor. Remember that, at the bottom of the movement, the knee will be bent at about 90°.

Level. Beginner, intermediate, and advanced

MUSCLE ACTIVATION + 1,5 –

CAUTION

Make sure you are in a stable position on three support points before starting to perform this exercise, and keep your back straight until your are done.

AVOID A COMMON MISTAKE

Avoid bending the knee excessively, because your leg would be bent at the rear of the thigh.

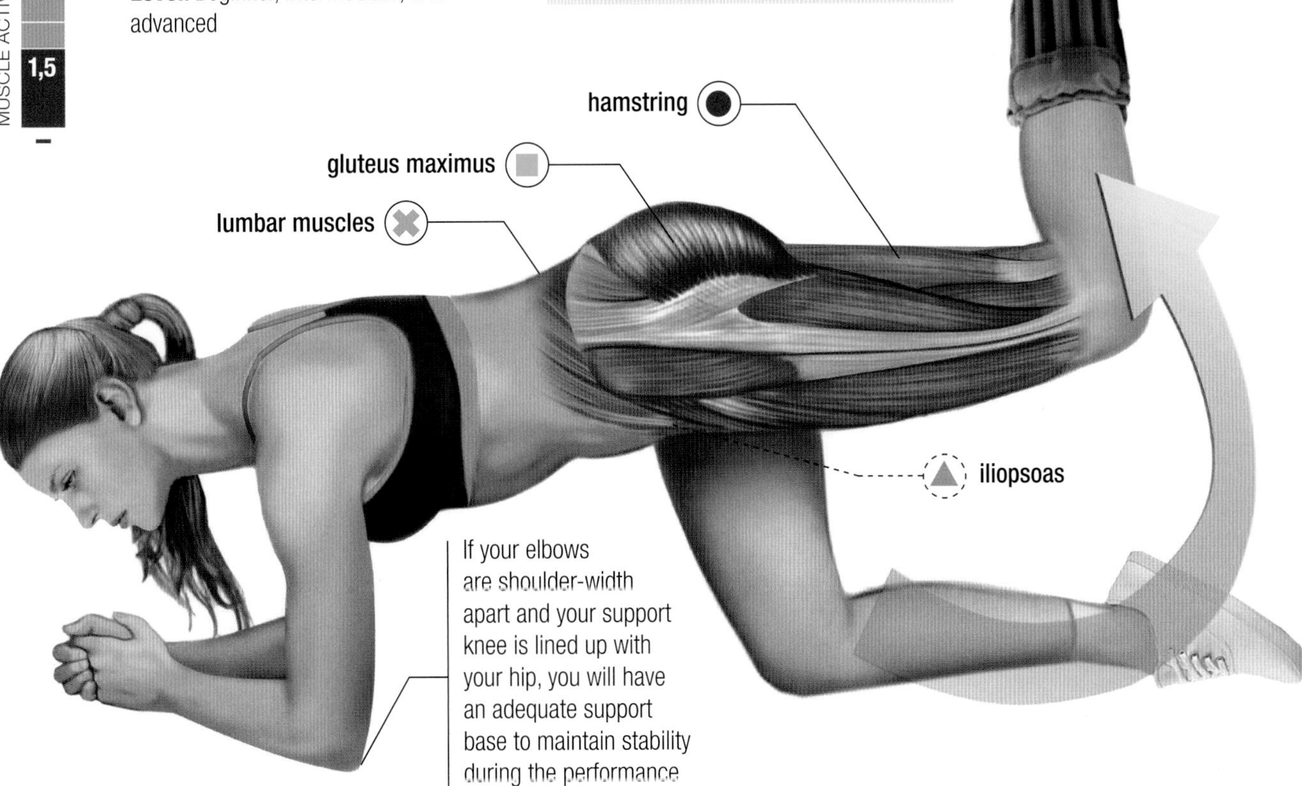

If your elbows are shoulder-width apart and your support knee is lined up with your hip, you will have an adequate support base to maintain stability during the performance of this exercise.

VARIANT

Gluteus maximus

Stationary Barbell Lunges

VARIANT

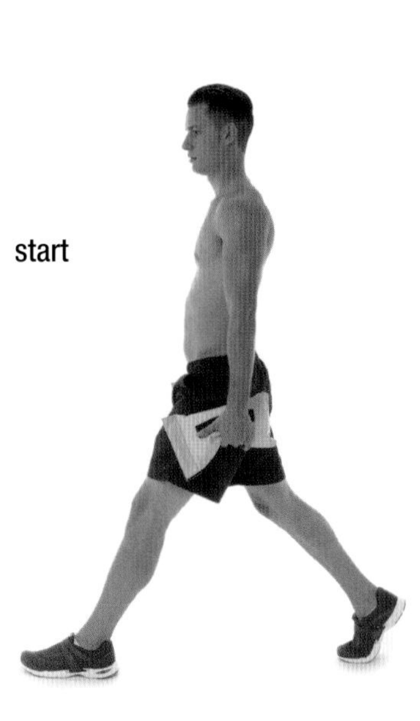

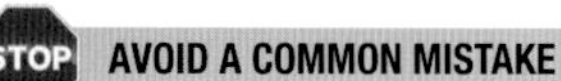

AVOID A COMMON MISTAKE

When you reach the bottom of this movement, if your knees are bent at an angle less than 90°, this means that your feet are too close together. If the angle is much greater than 90°, it means that your feet are too far apart. In both cases you need to correct your starting position.

CAUTION

Make sure your knees do not bend beyond 90° at any time during the movement.

Start. Place the bar across your shoulders, hold it with both hands, and place one foot ahead of the other. Keep your upper body perpendicular to the floor.

Technique. Lower your position until the knee of your trailing leg touches the floor, but keep it from hitting the floor too hard. Keep your back straight and perpendicular to the floor until you are in a lunge position. Then straighten your knees to return to the starting point.

Level. Beginner, intermediate, and advanced

Gluteus maximus

Supine Gluteals

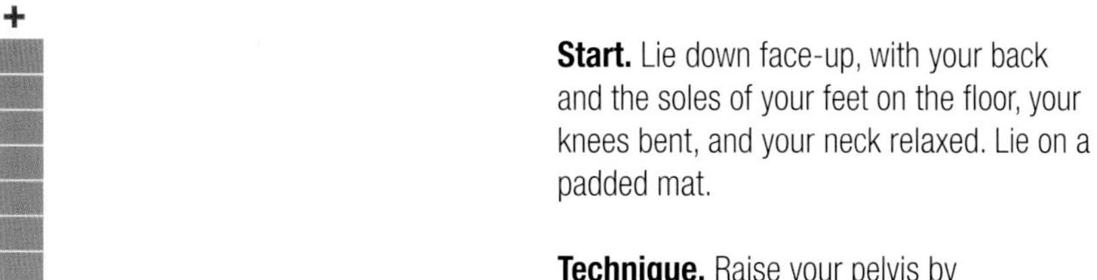

Start. Lie down face-up, with your back and the soles of your feet on the floor, your knees bent, and your neck relaxed. Lie on a padded mat.

Technique. Raise your pelvis by straightening your hips, until your upper body is aligned with your thighs, and return to the starting point. If the exercise is very easy for you, you can place a barbell plate, a dumbbell, or a weighted bag on your hips; this will increase the resistance.

Level. Beginner, intermediate, and advanced

AVOID A COMMON MISTAKE

Avoid straightening your cervical vertebrae by forcing things in this area.

CAUTION

Keep your back straight and aligned with your legs, without tipping to the side.

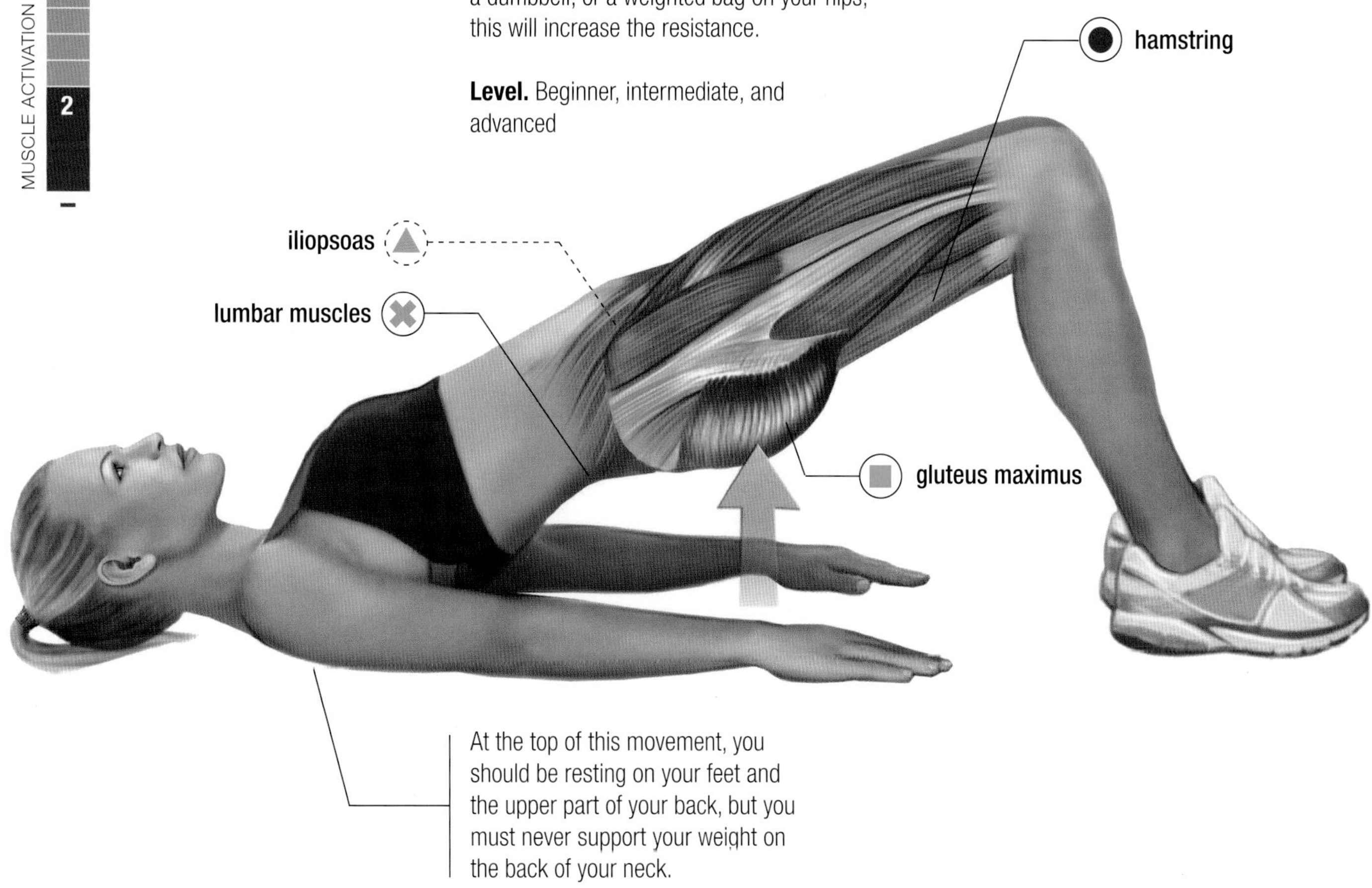

At the top of this movement, you should be resting on your feet and the upper part of your back, but you must never support your weight on the back of your neck.

VARIANT

Gluteus maximus

Hip Extensions

MUSCLE ACTIVATION + 4 –

Start. Take a position on a bench with the front of your upper body lying on it and your hands holding the end. Your hips and knees must be bent, and you can use ankle weights to increase the resistance.

Technique. Straighten your hips so that your thighs are aligned with your upper body, hold the position for a moment, and then return to the starting point. Keep the speed slow and consistent throughout the exercise.

Level. Beginner, intermediate, and advanced

AVOID A COMMON MISTAKE

Avoid taking a position too far back on the bench, because this could damage your spine in the lumbar region and reduce your stability.

CAUTION

Be careful on the downward movement; depending on your size and the height of the bench, your knees could hit the floor.

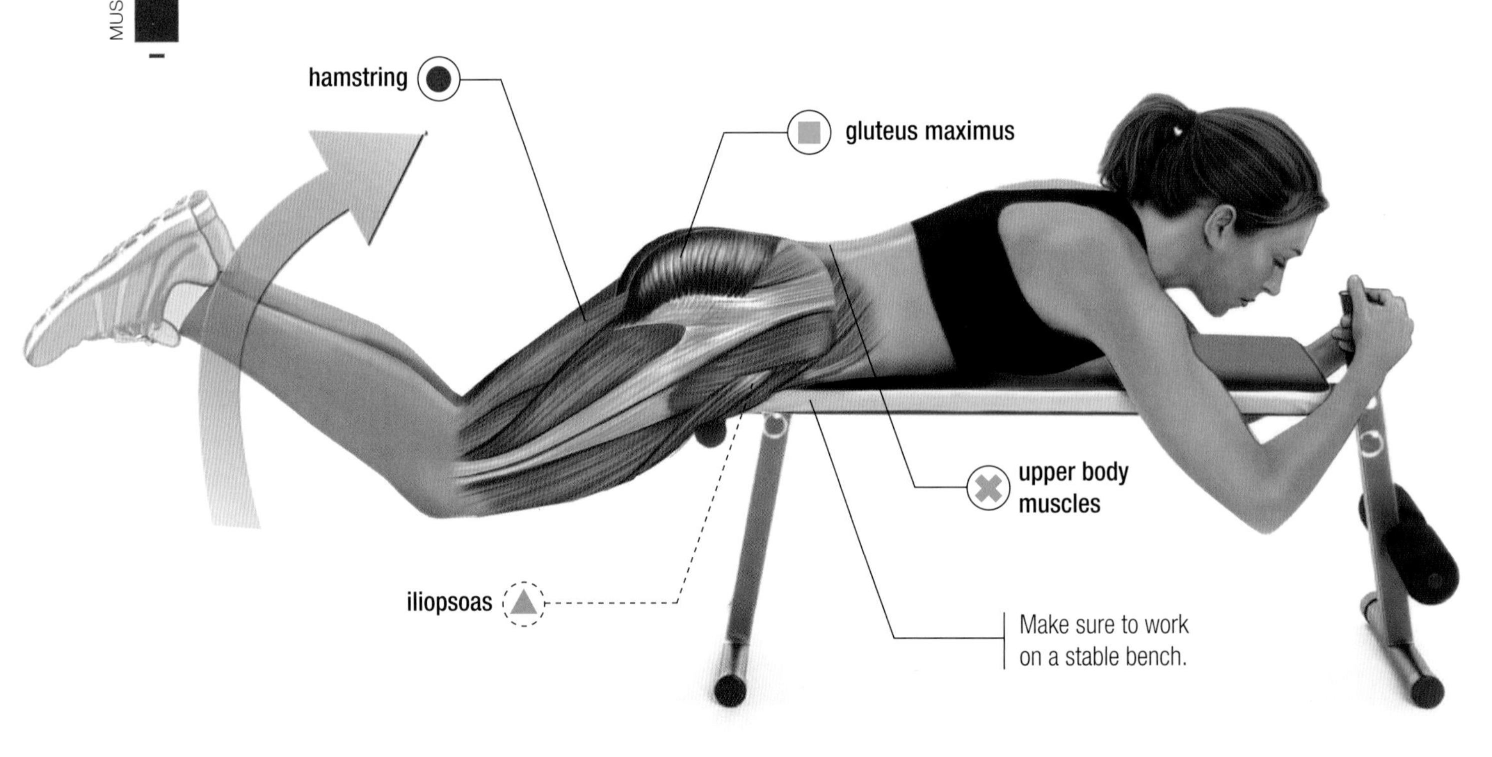

VARIANT

Raised Hip Abduction

Start. Lie down on your back on a padded mat, with your knees bent and your feet touching the floor. Raise your hips and support yourself on your feet and the top part of your back.

Technique. Move your knees apart and bring them back together while keeping your hips elevated. At the same time you are doing an isometric exercise for the gluteus maximus, you are doing isotonic work with the gluteus medius and minimus.

Level. Beginner, intermediate, and advanced

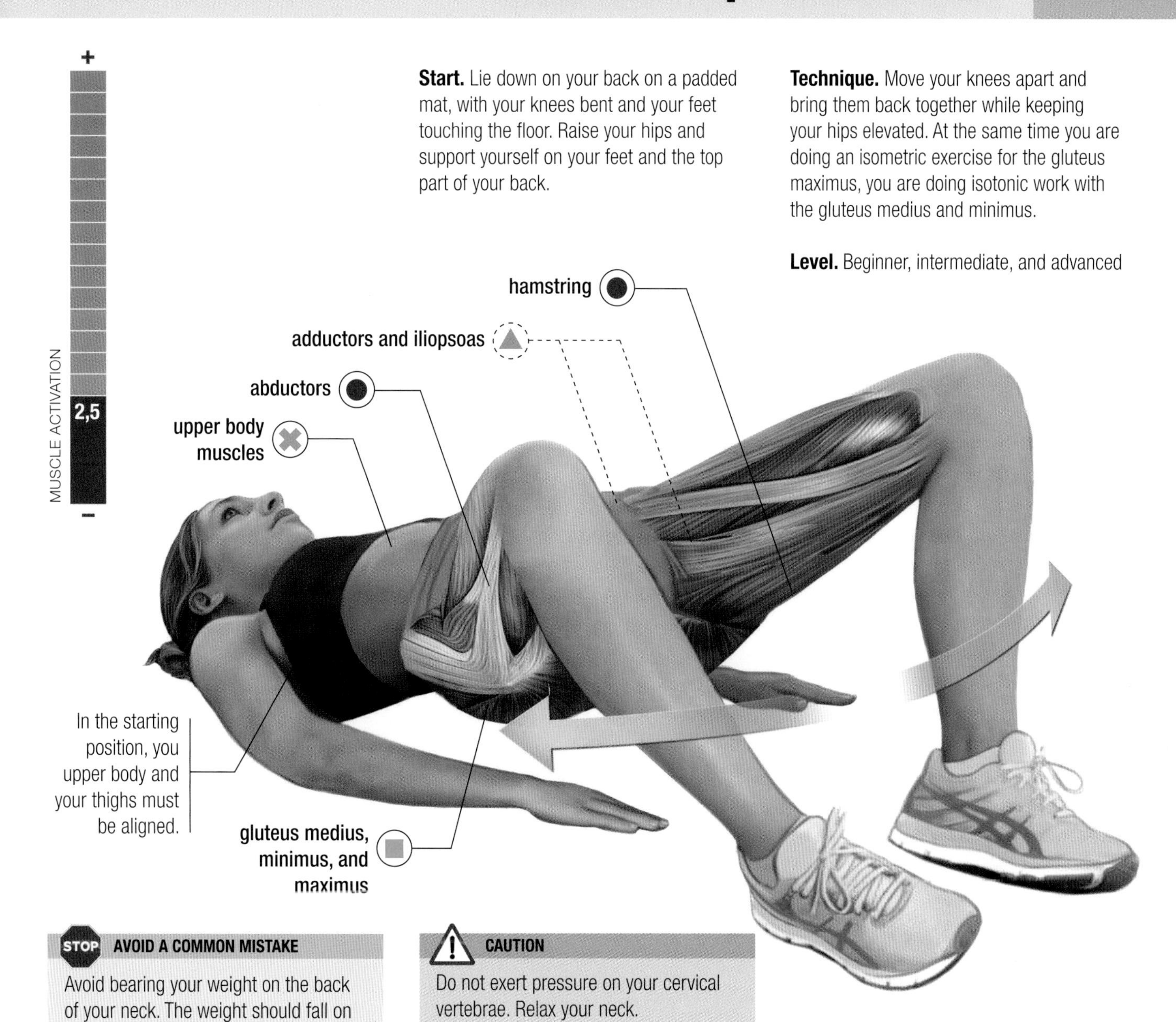

AVOID A COMMON MISTAKE

Avoid bearing your weight on the back of your neck. The weight should fall on the top part of your back and your feet.

CAUTION

Do not exert pressure on your cervical vertebrae. Relax your neck.

VARIANT

Abdominals

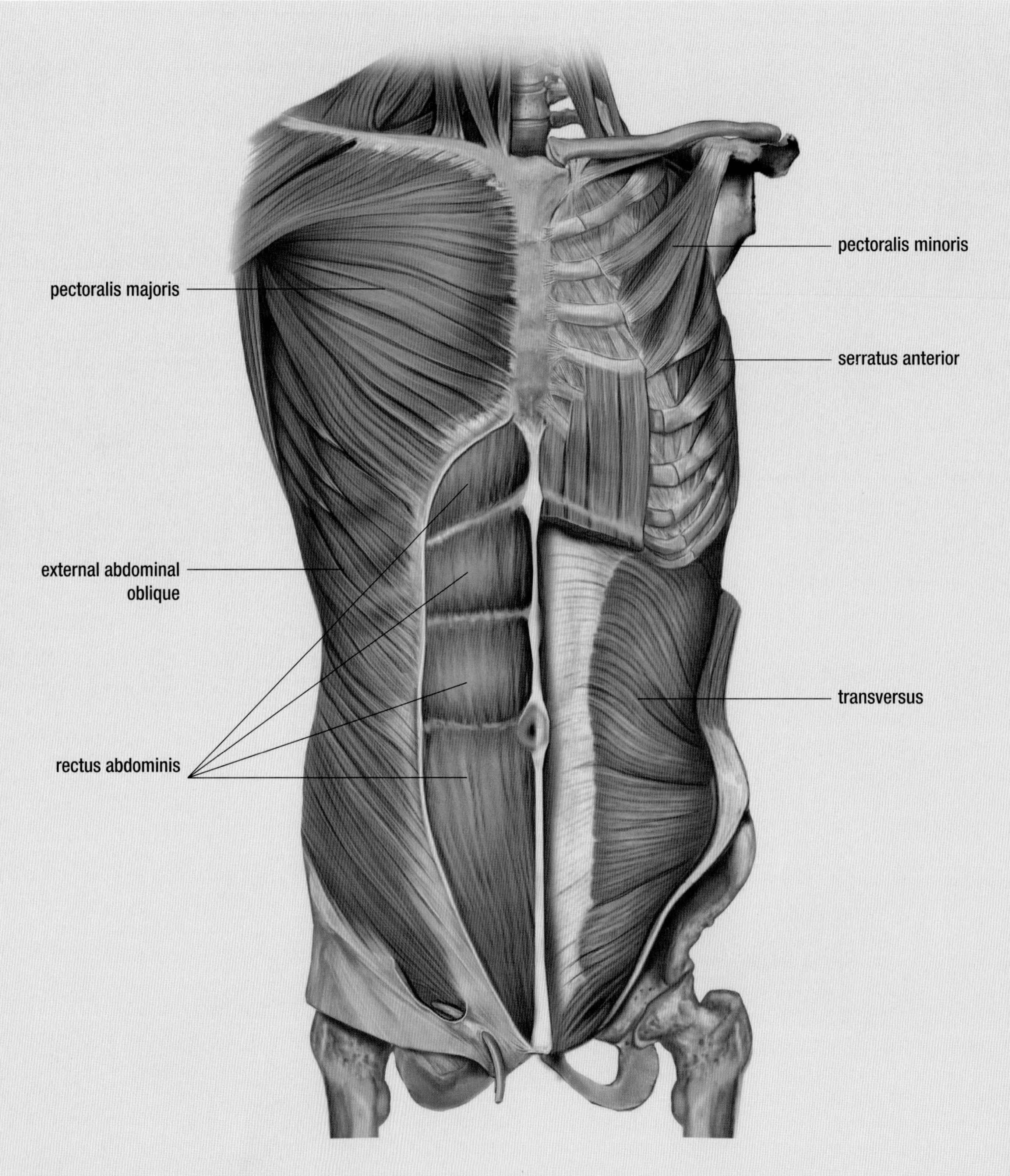
pectoralis minoris
pectoralis majoris
serratus anterior
external abdominal oblique
transversus
rectus abdominis

The abdominal muscles occupy the area from the lower part of the thorax to the upper part of the pelvis and cover the abdominal cavity. These muscles are of great importance in maintaining proper posture and in holding and protecting the internal organs. For bodybuilders, they also have a clearly aesthetic appeal, especially in the case of the rectus abdominis. Still, they are not very thick, and, for them to be visible, they not only have to be worked thoroughly, but also the amount of accumulated fat on them must be much reduced. People can have very powerful abdominals, but without showing them (because of fat accumulation).

Rectus abdominis. This muscle arises in the pubis, and it inserts at the fifth, sixth, and seventh ribs and the sternum. Its function is to bend the upper body, and it is the most visible of the abdominal muscles. It is also the best-known one, because of its arrangement in the shape of packets or "squares." It is an important muscle, because of its active use in combat sports on the floor, such as judo and Greco–Roman wrestling, and because of its passive use for protection in sports, such as boxing. It is also used in a stabilizing function and for breathing in most, if not all, sports.

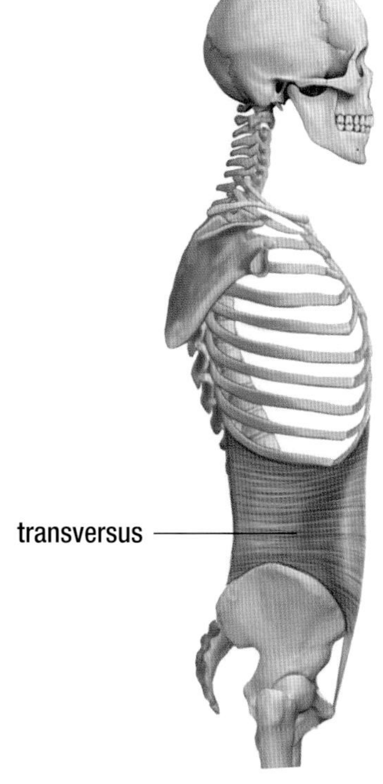

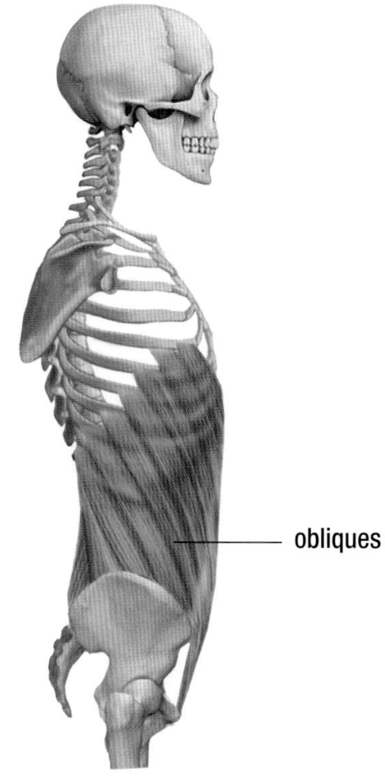

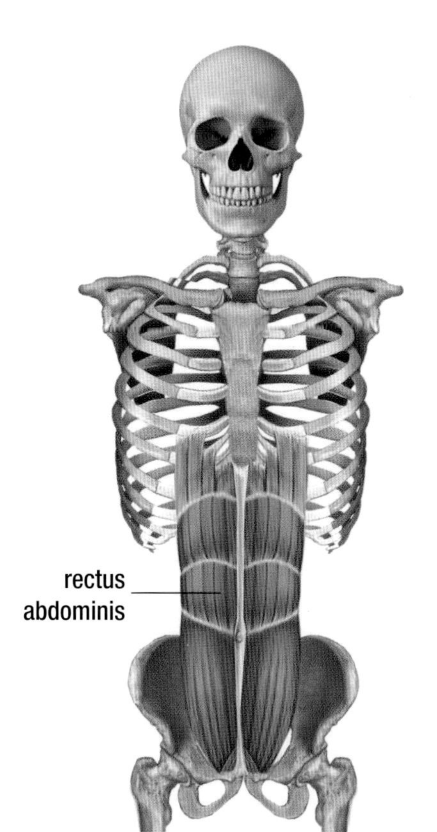

Transversus. This muscle arises at the inguinal ligament, the iliac crest, the thoracolumbar fascia, and the costal cartilage of the seventh through the twelfth ribs, and it inserts at the linea alba and the pubic crest. The main functions of this muscle are to contain and support the internal organs, and forceful exhaling. As a result, it is vitally important in all sports for controlling breathing, especially in the various swimming disciplines, where breathing is crucial. It is also used in combat sports in which forceful exhaling is required at the instant of striking or falling.

Obliques. There are major, or external, and minor, or internal, oblique muscles. The major oblique arises at the fifth through the twelfth ribs, and it inserts at the iliac crest, the thoracolumbar fascia, the linea alba, and the pubis. The minor oblique originates at the iliac crest, the thoracolumbar fascia, and the inguinal ligament, and it inserts at the ninth through the twelfth ribs, the aponeurosis of the transversus, the inguinal ligament, the linea alba, and the cartilage of ribs seven through nine. The main function of both muscles is rotation of the upper body, but they also contribute to bending it, and they are used significantly for hitting in sports that use implements (rackets, sticks, clubs, bats, and so on.). As a result, they are used in sports such as tennis, paddleball, hockey, baseball, badminton, squash, paddleball, jai alai, cricket, and polo.

Rectus abdominis

Cable Crunches

MUSCLE ACTIVATION + 6 −

Start. Take a kneeling position with your back to the machine and hold the grip with both hands, one on each side of your head. The pulley must be anchored high.

Technique. Gently move your upper body forward and downward. Remember that, if you seek to bend your upper body to work the rectus abdominis, the range of motion will be between 2–4 inches (5–10 cm).

Level. Intermediate and advanced

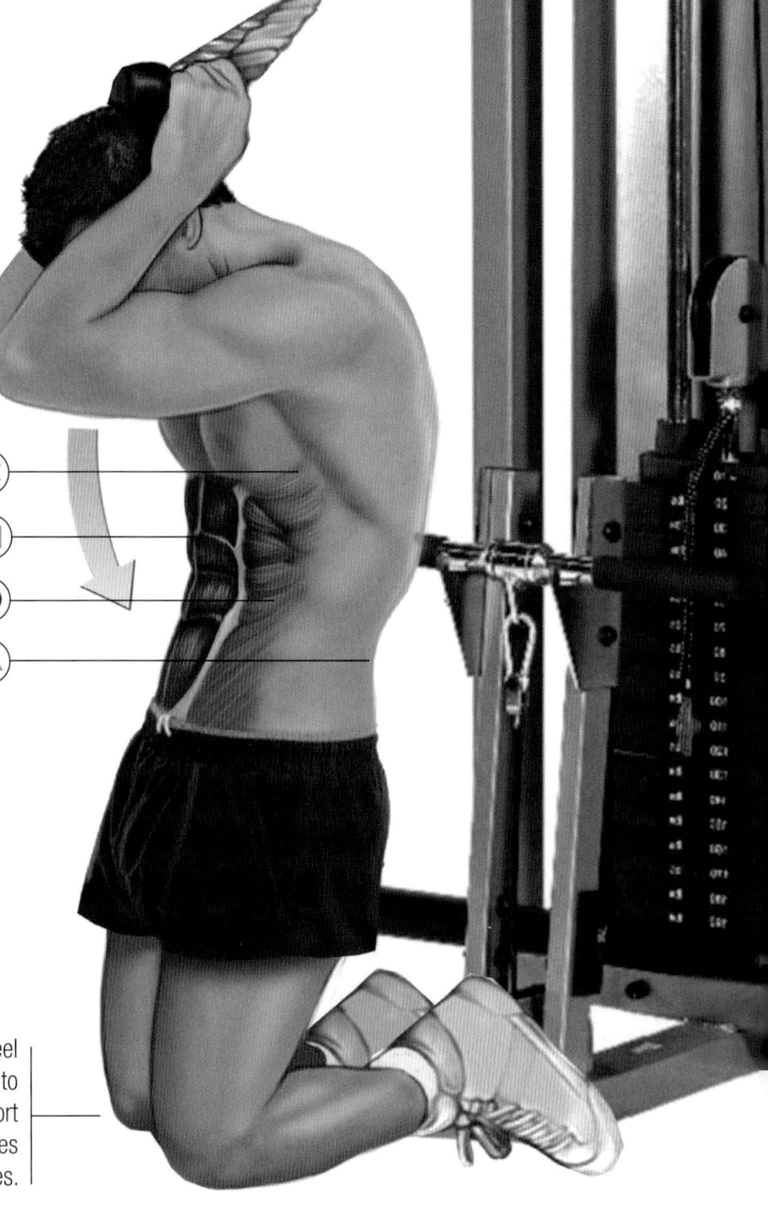

AVOID A COMMON MISTAKE

Avoid bending your hips, because what you want is a bend in your upper body, and that is where you will achieve quality work.

CAUTION

Avoid bringing your chest too close to the floor, because that would give too much priority to the iliopsoas at the expense of the rectus abdominis.

You can kneel on a mat to avoid discomfort in your knees and ankles.

VARIANT

start

finish

110 / ANATOMY AND BODYBUILDING

Rectus abdominis

Upper Body Cable Flex

Start. Lie down on a mat with your head toward the machine and your feet pointing away from it. The pulley must be adjusted to a low position, and you should hold the rope or the single grip with both hands. Keep your elbows bent at about 90° and hold your hands about a hand's length from your forehead.

Technique. Lift your chest slightly by bending your upper body, while keeping your arms in the same position. Then go back down until your entire back is resting on the mat.

Level. Intermediate and advanced

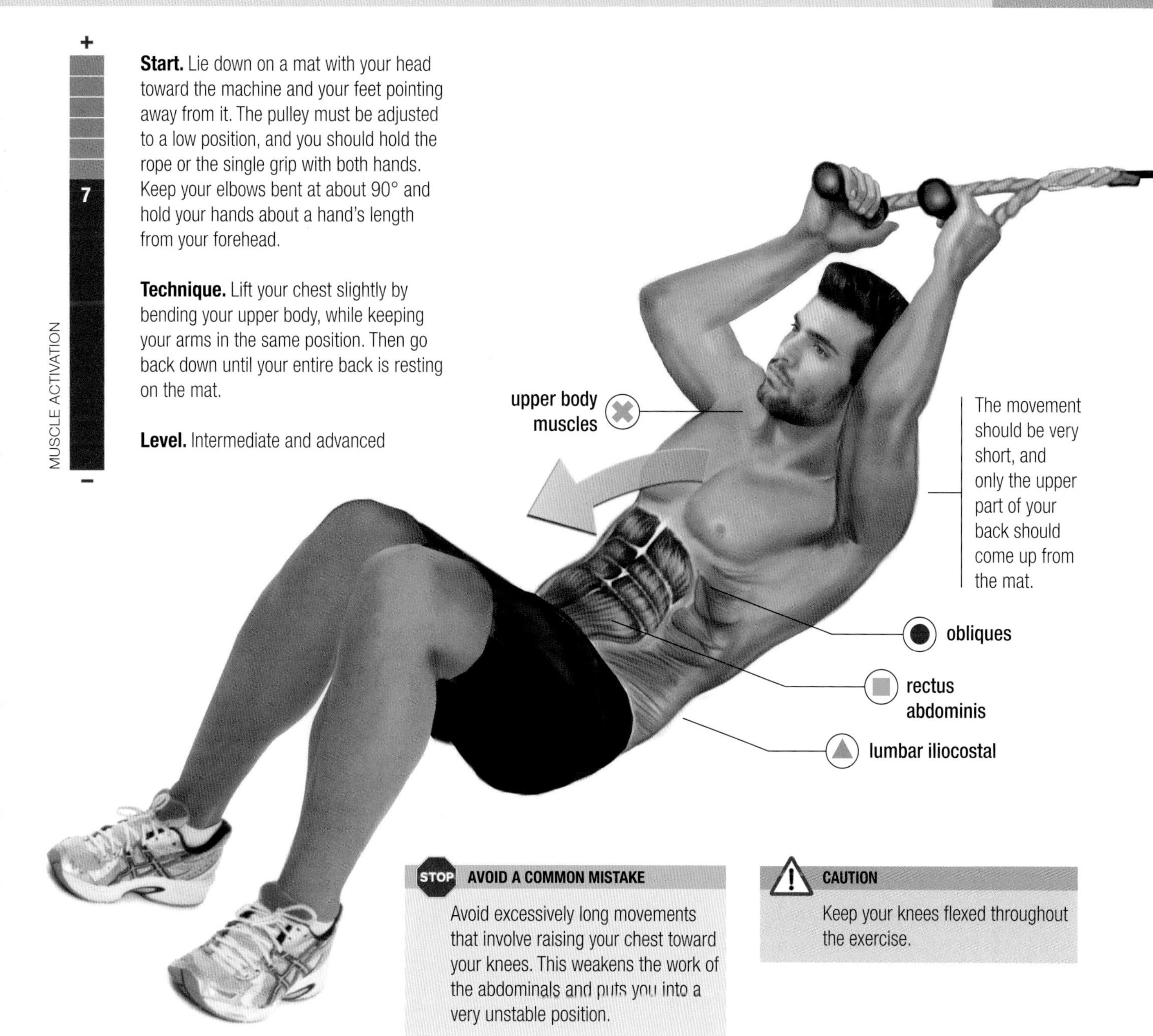
+
7
MUSCLE ACTIVATION
–
upper body muscles
The movement should be very short, and only the upper part of your back should come up from the mat.
obliques
rectus abdominis
lumbar iliocostal

AVOID A COMMON MISTAKE

Avoid excessively long movements that involve raising your chest toward your knees. This weakens the work of the abdominals and puts you into a very unstable position.

CAUTION

Keep your knees flexed throughout the exercise.

VARIANT

start
finish

Rectus abdominis

Upper Body Flex with Dumbbells

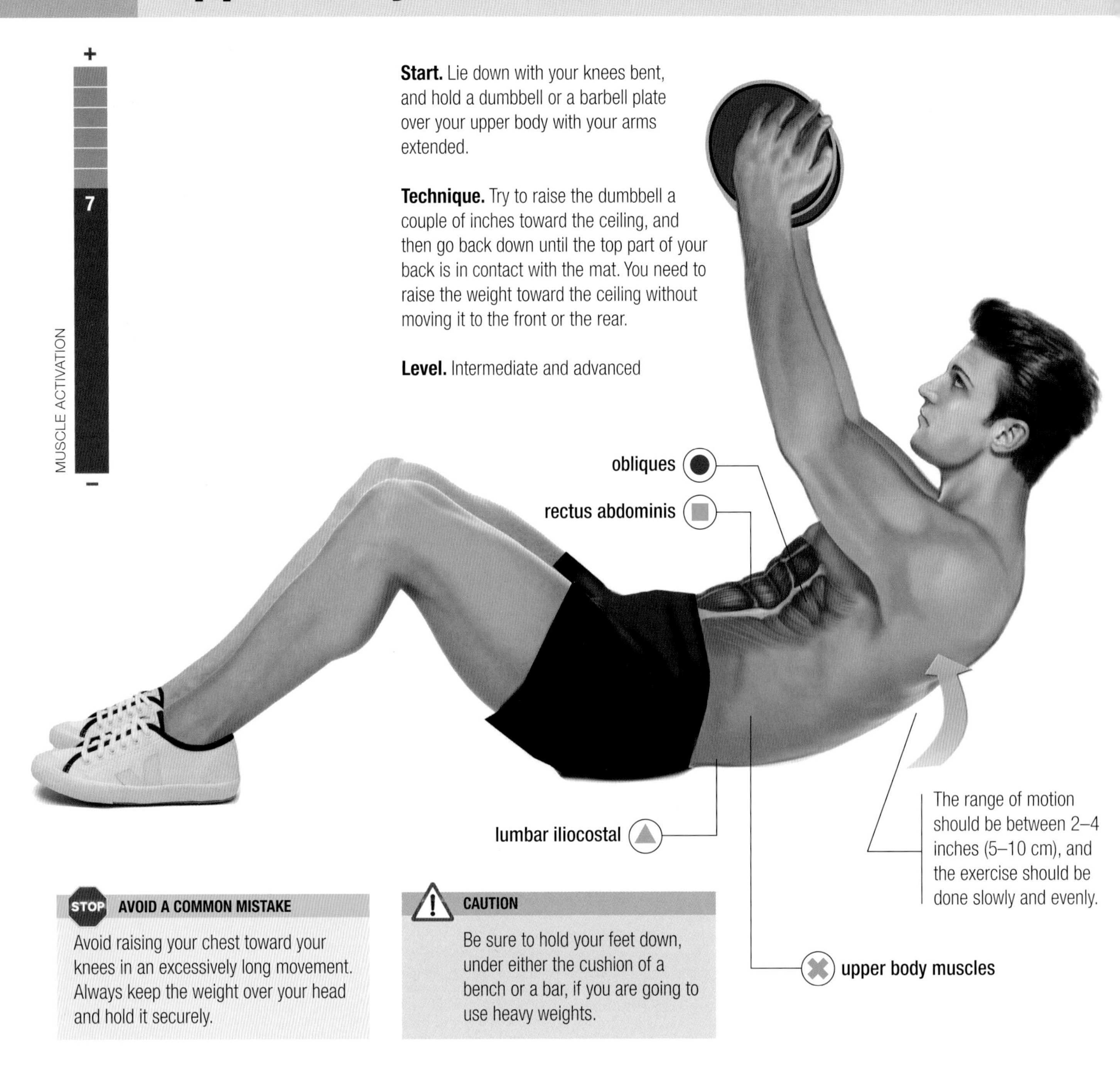
+
7
MUSCLE ACTIVATION
–
obliques
rectus abdominis
lumbar iliocostal
upper body muscles
The range of motion should be between 2–4 inches (5–10 cm), and the exercise should be done slowly and evenly.

Start. Lie down with your knees bent, and hold a dumbbell or a barbell plate over your upper body with your arms extended.

Technique. Try to raise the dumbbell a couple of inches toward the ceiling, and then go back down until the top part of your back is in contact with the mat. You need to raise the weight toward the ceiling without moving it to the front or the rear.

Level. Intermediate and advanced

AVOID A COMMON MISTAKE

Avoid raising your chest toward your knees in an excessively long movement. Always keep the weight over your head and hold it securely.

CAUTION

Be sure to hold your feet down, under either the cushion of a bench or a bar, if you are going to use heavy weights.

VARIANT

start
finish

Crunches on the Machine

+
5
MUSCLE ACTIVATION
–

Start. Sit down on the abdominal machine, with the cushion pressed against your chest and your feet secured on the footrest. Keep your hands on the chest pad for safety.

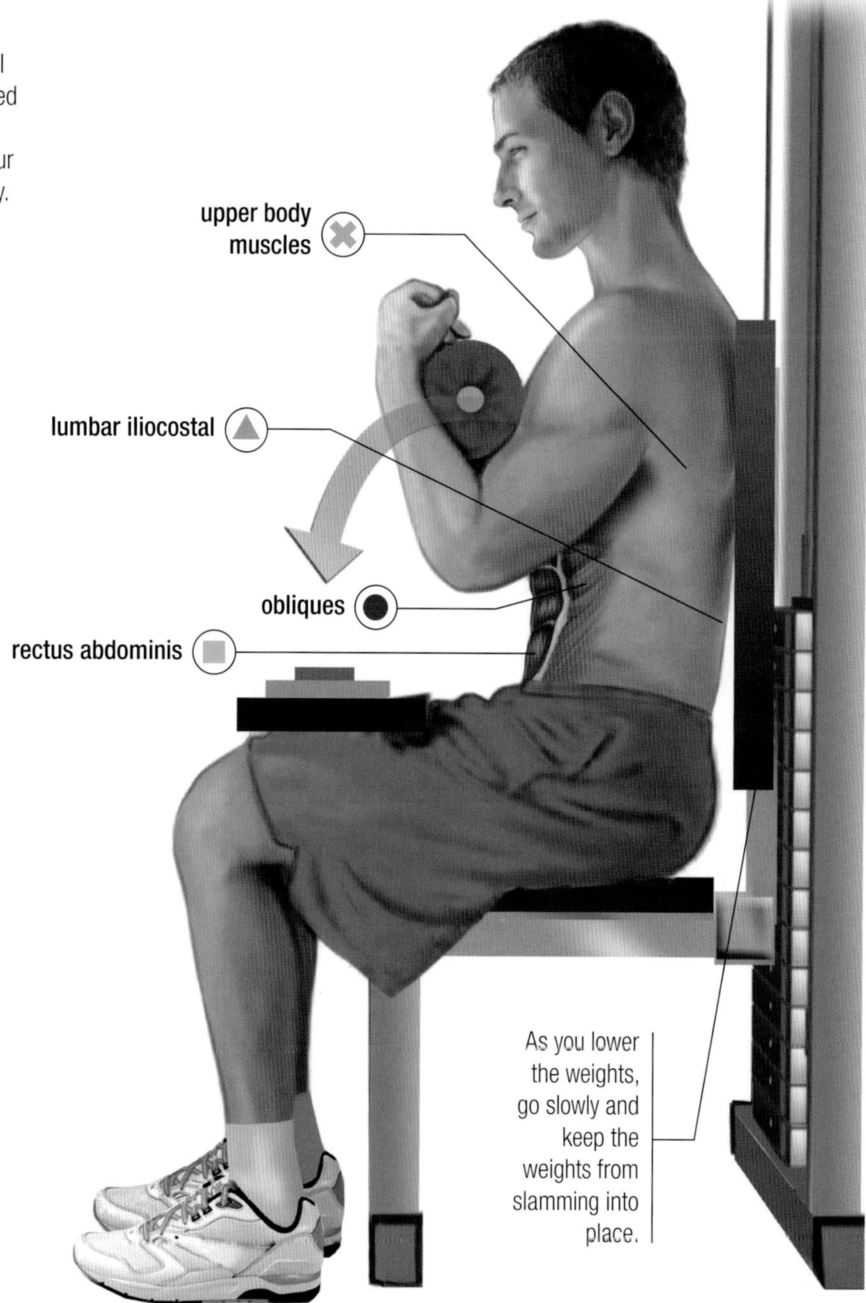
upper body muscles
lumbar iliocostal
obliques
rectus abdominis
As you lower the weights, go slowly and keep the weights from slamming into place.

Technique. Bend the upper body slightly and lower your chest, and then return to the starting point, keeping the weights from slamming together. The movement is short, and it should be done slowly and without jerking.

Level. Beginner, intermediate, and advanced

AVOID A COMMON MISTAKE

Avoid bending at your hips instead of bending your upper body. You will notice, that, if you are doing it wrong, you will feel excessive tension in your legs.

CAUTION

Avoid jerking and sudden movements when doing this exercise, because they could damage your back.

VARIANT

start

finish

Rectus abdominis

Upper Body Flex with Barbell Plate and Arms Extended

MUSCLE ACTIVATION
+
7
–

AVOID A COMMON MISTAKE

Avoid moving the plate closer to the knees. The shoulders and elbows should remain stationary. You should lift the plate only by bending your upper body.

CAUTION

Don't do very long movements. Remember, this movement is only 2–4 inches (5–10 cm) long. Start without using a weight, or with very light weights, and increase the weight gradually.

Start. Lie down on your back on a mat and hold a barbell plate with both hands. Keep your elbows almost completely straight and in line with your upper body.

Technique. Bend your upper body slightly by raising the barbell plate toward the ceiling and keeping your arms almost totally straight. Slowly return to the starting point, maintaining control.

Level. Intermediate and advanced

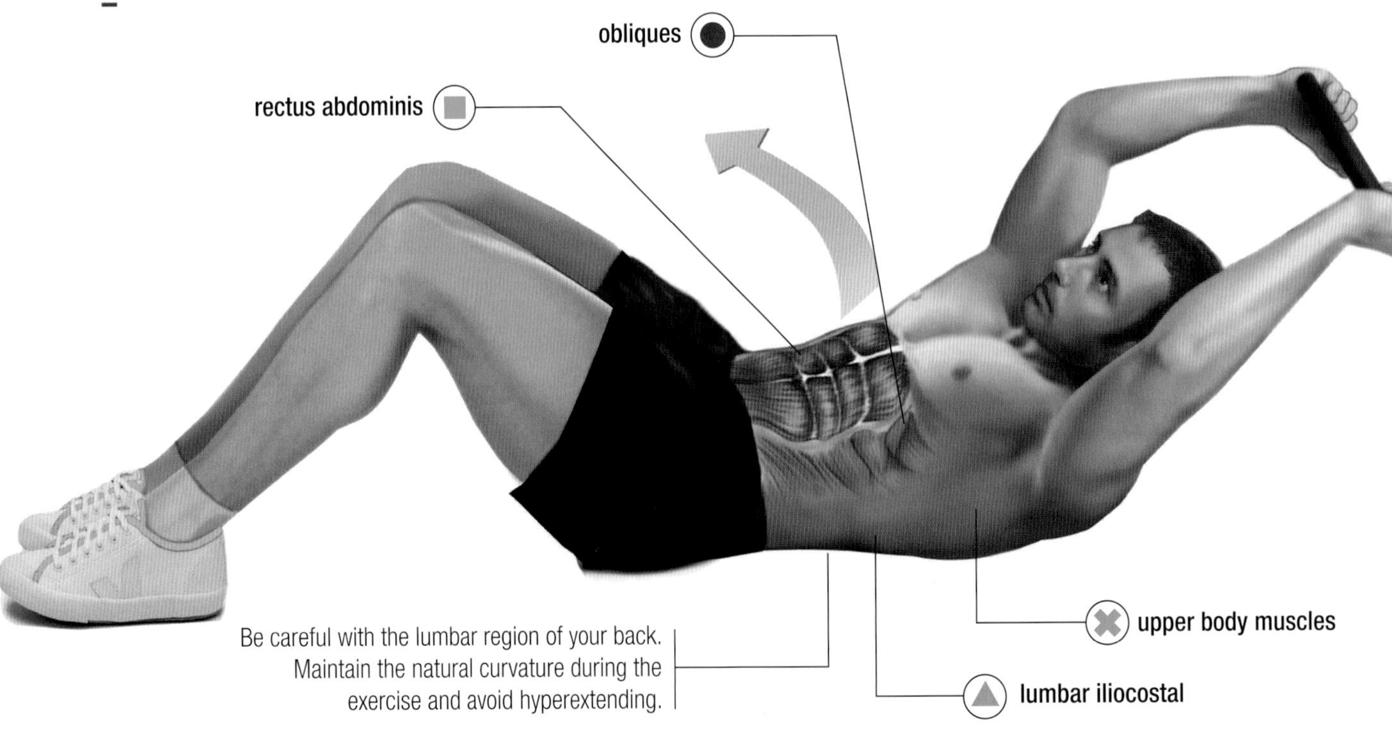

Be careful with the lumbar region of your back. Maintain the natural curvature during the exercise and avoid hyperextending.

VARIANT

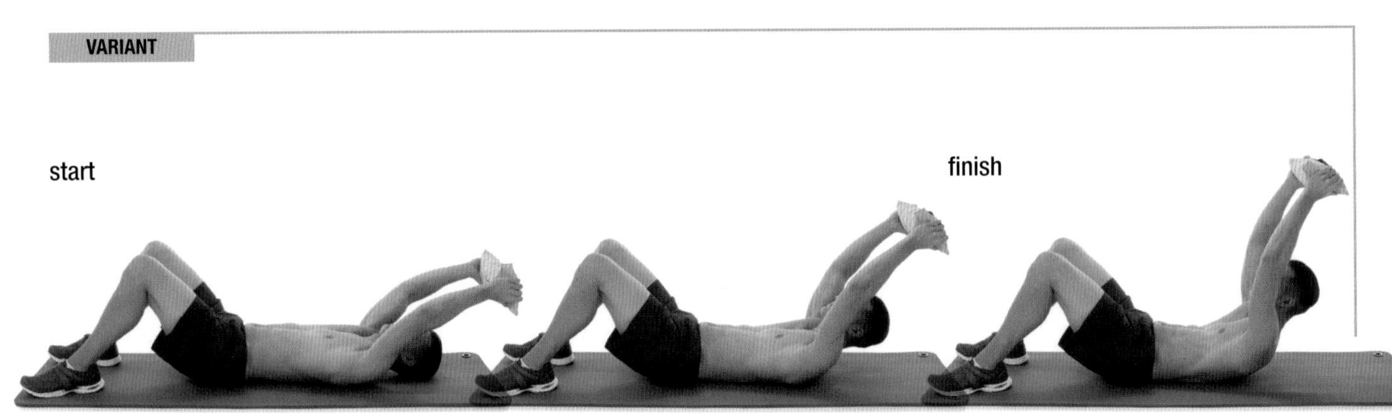

Abdominal Compression

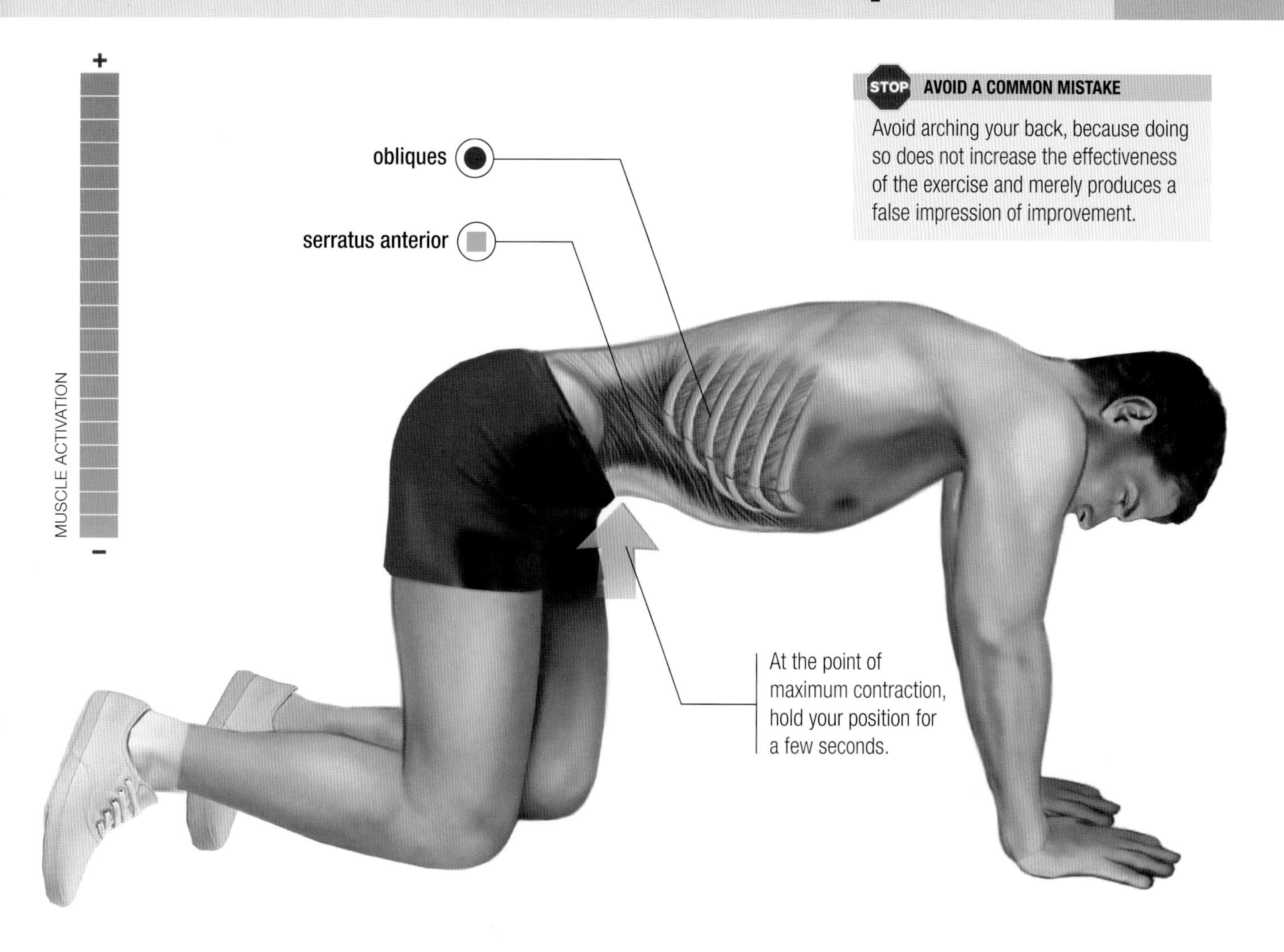

STOP AVOID A COMMON MISTAKE

Avoid arching your back, because doing so does not increase the effectiveness of the exercise and merely produces a false impression of improvement.

VARIANT

Start. Get down on all fours on a mat as shown in the illustration, and keep your hands in line with your feet so your position is steady.

Technique. Suck in your abdomen, trying to reduce the distance from your navel to your spine as much as possible. Then relax your abdomen and return to the starting point. Don't forget to do the movement in a slow, controlled manner.

Level. Beginner, intermediate, and advanced

CAUTION

Keep your spine in the same position throughout the exercise.

Extended Obliques

MUSCLE ACTIVATION: 5

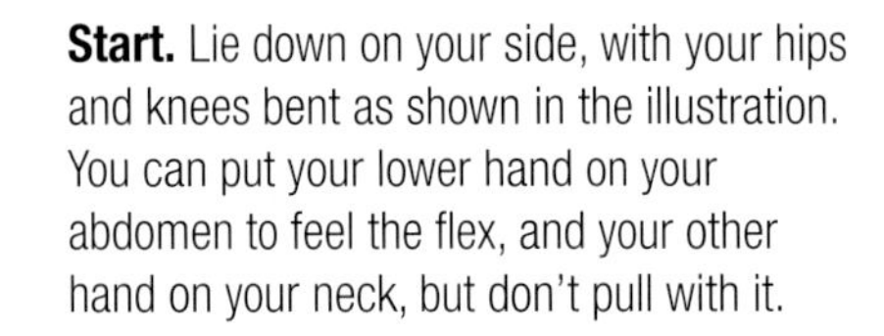

Start. Lie down on your side, with your hips and knees bent as shown in the illustration. You can put your lower hand on your abdomen to feel the flex, and your other hand on your neck, but don't pull with it.

Technique. Bend your upper body sideways and try to reduce the distance between your underarm and your hip. The movement should be short, and it should be done slowly and under control to optimize the results from the exercise.

Level. Beginner, intermediate, and advanced

AVOID A COMMON MISTAKE

Avoid pulling your head or neck with the hand that is placed on your neck. It must merely rest on the back of your neck, and it must not pull or apply any pressure.

CAUTION

Remember that in abdominal exercises the motion is always very short, and this is no exception. If you perform an excessively long motion, your technique will deteriorate, the results will be minimal, and you may end up in an unstable support position with precarious balance.

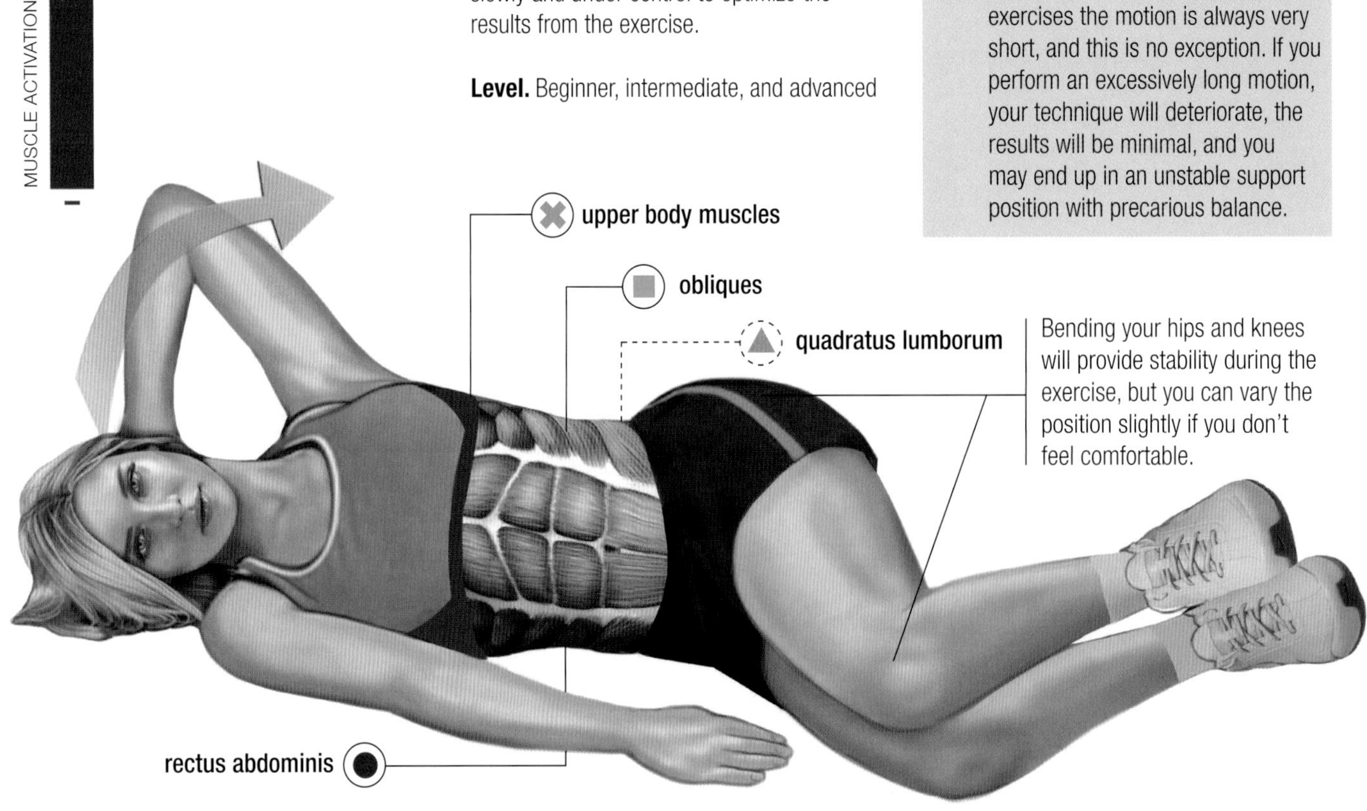

Bending your hips and knees will provide stability during the exercise, but you can vary the position slightly if you don't feel comfortable.

VARIANT

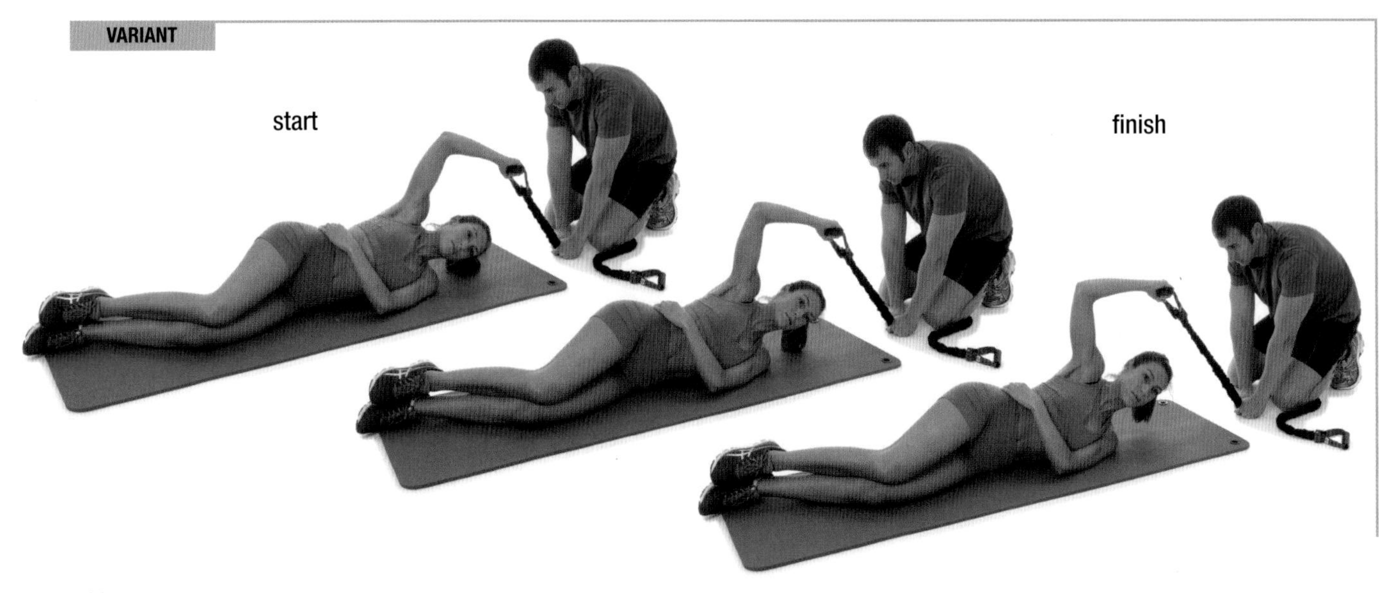

Bent-leg Obliques

Start. Lie down on a mat and put each hand on the opposite shoulder. You must keep your knees bent, one foot on the floor, and the other one crossed, resting your ankle on the opposite knee, as shown in the illustration.

Technique. Try to bring your elbow to the opposite knee by bending and rotating your upper body. Remember that the knee toward which you move your elbow is the one on the crossed leg. Once you do the exercise with one leg, repeat the same process using the opposite side.

Level. Beginner, intermediate, and advanced

AVOID A COMMON MISTAKE

Avoid moving your knee toward your chest instead of raising your chest toward your knee. Even though it appears that the distance is being reduced in the same way, the two movements do not involve the same muscles.

CAUTION

Avoid jerking motions and letting your upper body fall back in the negative phase of the exercise. Even though you have a mat underneath your back, sudden drops and an impact with the floor can produce a negative effect on your vertebrae.

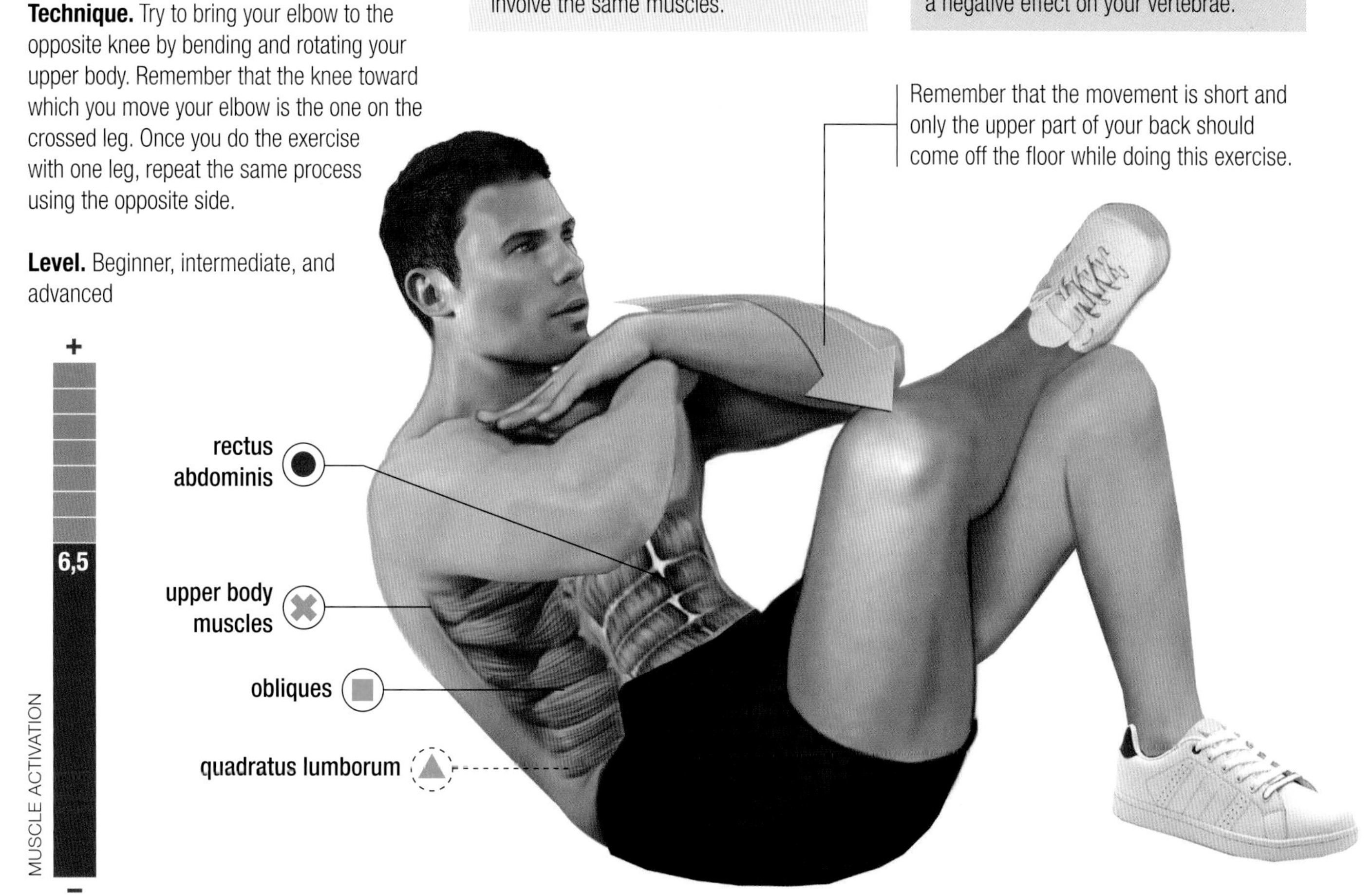
Remember that the movement is short and only the upper part of your back should come off the floor while doing this exercise.
+
6,5
MUSCLE ACTIVATION
–
rectus abdominis
upper body muscles
obliques
quadratus lumborum

VARIANT

start
finish

Obliques with Pulley

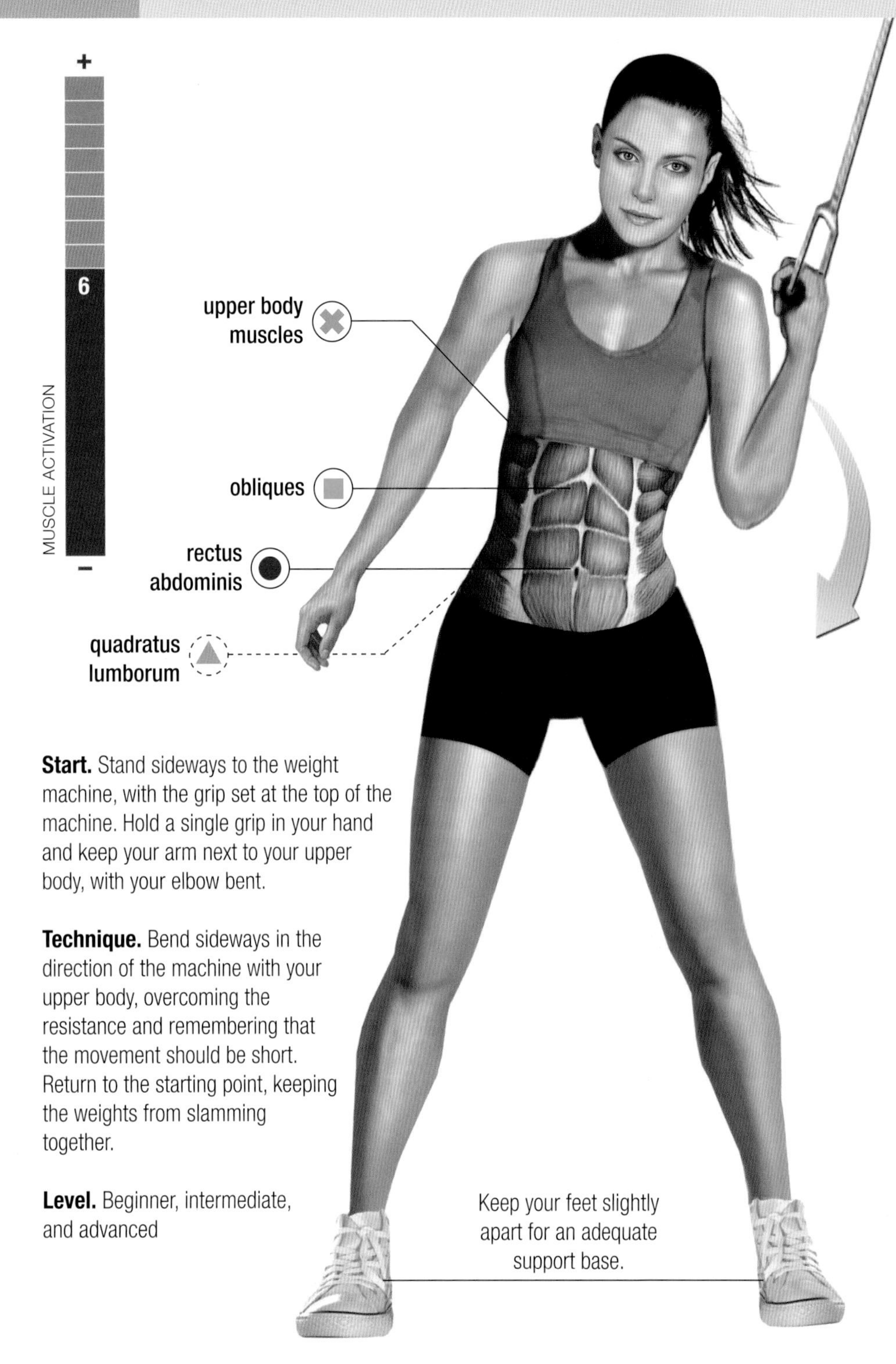

Start. Stand sideways to the weight machine, with the grip set at the top of the machine. Hold a single grip in your hand and keep your arm next to your upper body, with your elbow bent.

Technique. Bend sideways in the direction of the machine with your upper body, overcoming the resistance and remembering that the movement should be short. Return to the starting point, keeping the weights from slamming together.

Level. Beginner, intermediate, and advanced

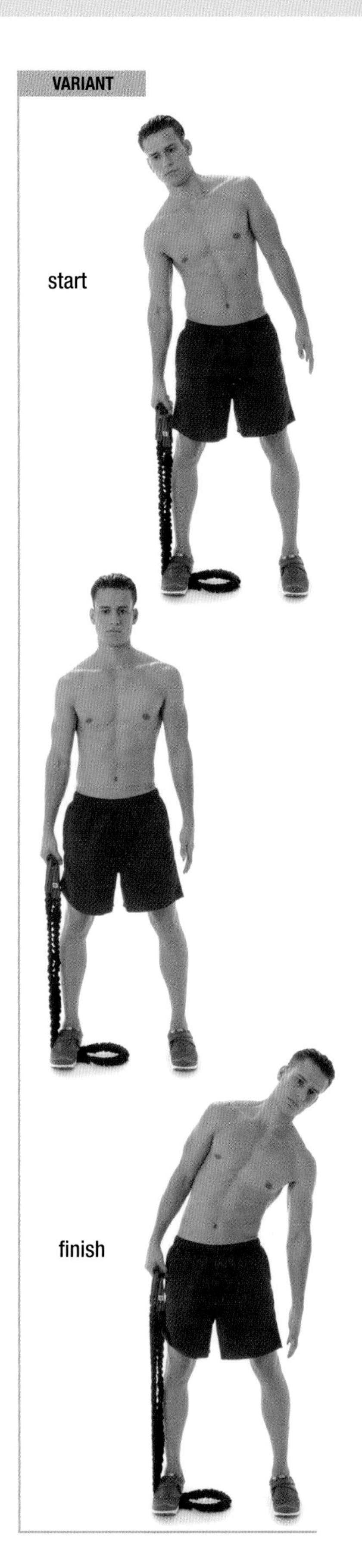

AVOID A COMMON MISTAKE

Avoid pulling the cable downward by straightening your elbow instead of bending your upper body to the side. This is a common mistake among new bodybuilders who try to lengthen the range of motion of this exercise more than they should.

CAUTION

Avoid moving your shoulder and elbow, because they must remain stationary throughout the exercise for it to be effective.

Dumbbell Twists

Start. Sit with your knees bent at 90° and your feet on the floor. Keep your upper body inclined slightly to the rear. Hold a dumbbell, barbell plate, or a medicine ball in both hands. Keep your arms nearly straight toward the front.

Technique. Move the weight toward one side, above your legs, by rotating your upper body. Then do the movement in the opposite direction. Remember to perform the exercise in a controlled manner.

Level. Intermediate and advanced

+
6
MUSCLE ACTIVATION
–
upper body muscles
rectus abdominis
obliques
quadratus lumborum
Do the movement slowly, but not too slowly, because you perform most of the work when you stop the inertia of the movement in one direction and start in the other.
VARIANT

STOP AVOID A COMMON MISTAKE

Avoid placing your feet very close together or very close to your buttocks, because this reduces your stability in an exercise that is inherently unstable.

CAUTION

Start working with light weights, because this exercise is technically complex and, at first, you may find it difficult to maintain a balanced position.

Training Level

Beginner

Athletes who are new to the practice of bodybuilding need to know that, at the outset, it may be difficult for them to do certain exercises correctly, especially ones involving free weights, particularly if those exercises involve several joints or working angles that are unstable. This is the case, for example, with the declined bench press. Also, at first, it may be difficult for novice bodybuilders to identify the isolation of a given muscle. So, for beginners, it's a good idea to start working on weight machines that guide movement and which reduce the number of joints involved, in order not to interfere with achieving good results. This can help beginners identify sensations and to perfect technical execution—in other words, to thoroughly know their own bodies and the sensations that they produce.

Now this book will present two appropriate workouts for beginners, which can be done at a gym, as well as alternatives for places that don't have specific equipment.

It is important to remember that a beginning workout can be used for a very

Workout A / Beginning Level

1 Leg Extension

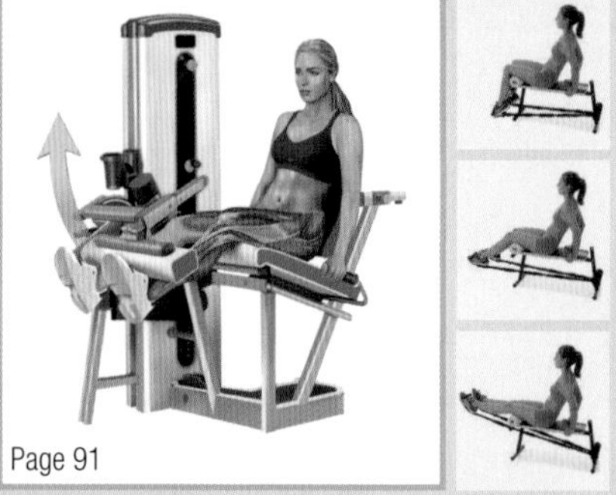

Page 91

2 Hamstring Curls

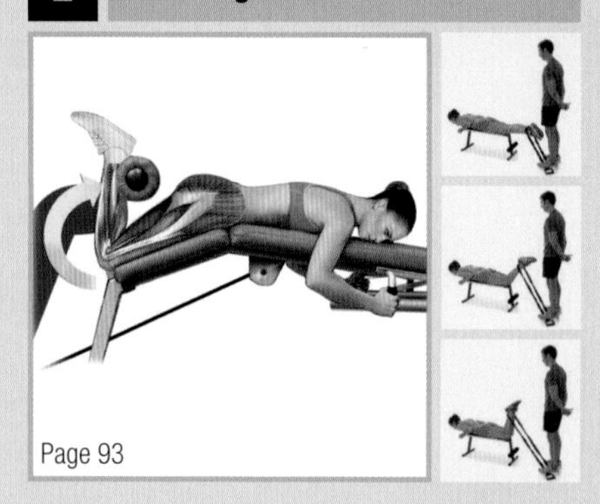

Page 93

3 Gluteal Raises on all Fours, with Ankle Weight

Page 103

4 Soleus on the Machine

Page 97

5 Peck-deck

Page 31

6 Rowing on the Machine

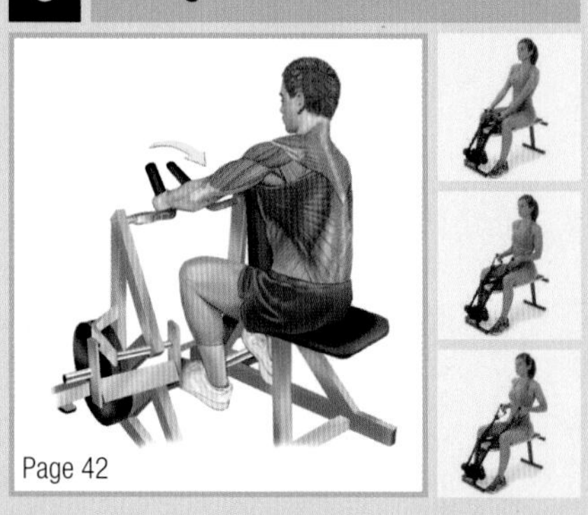

Page 42

7 Lateral Raises on the Machine

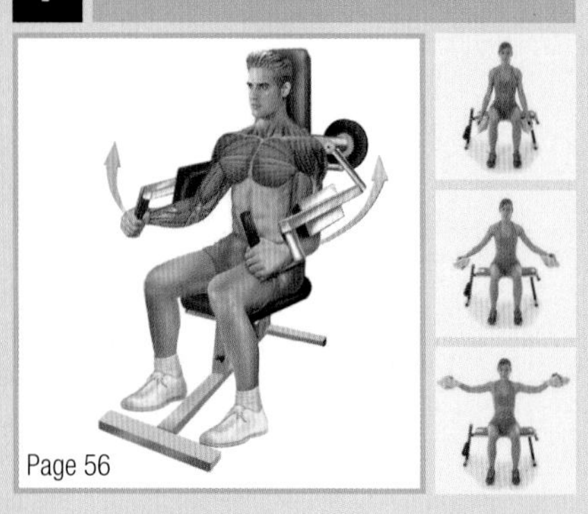

Page 56

8 Regular Pull-down

Page 75

9 Crunches on the Machine

Page 113

long time, retaining effectiveness and creating improvements in physique. A beginning workout is not synonymous with low performance and limited improvement. To make gains, this book recommends using the training log presented in the introduction. That's where you write down your progress, the weights used in each exercise, and the repetitions done to the point of muscle failure. With these data, you will have a point of reference to try to beat in subsequent training sessions. If you bench press 130 pounds (60 kg) 10 times, the goal will be 11 reps in subsequent workouts. When it's possible to do 12 within the time limit of 90 seconds, this is the time to add weight, which will require reducing the number of repetitions, thereby starting a new cycle of progress and improvement. This way, the workouts for beginners, intermediates, and advanced bodybuilders can all be used for a long time before you experience stagnation.

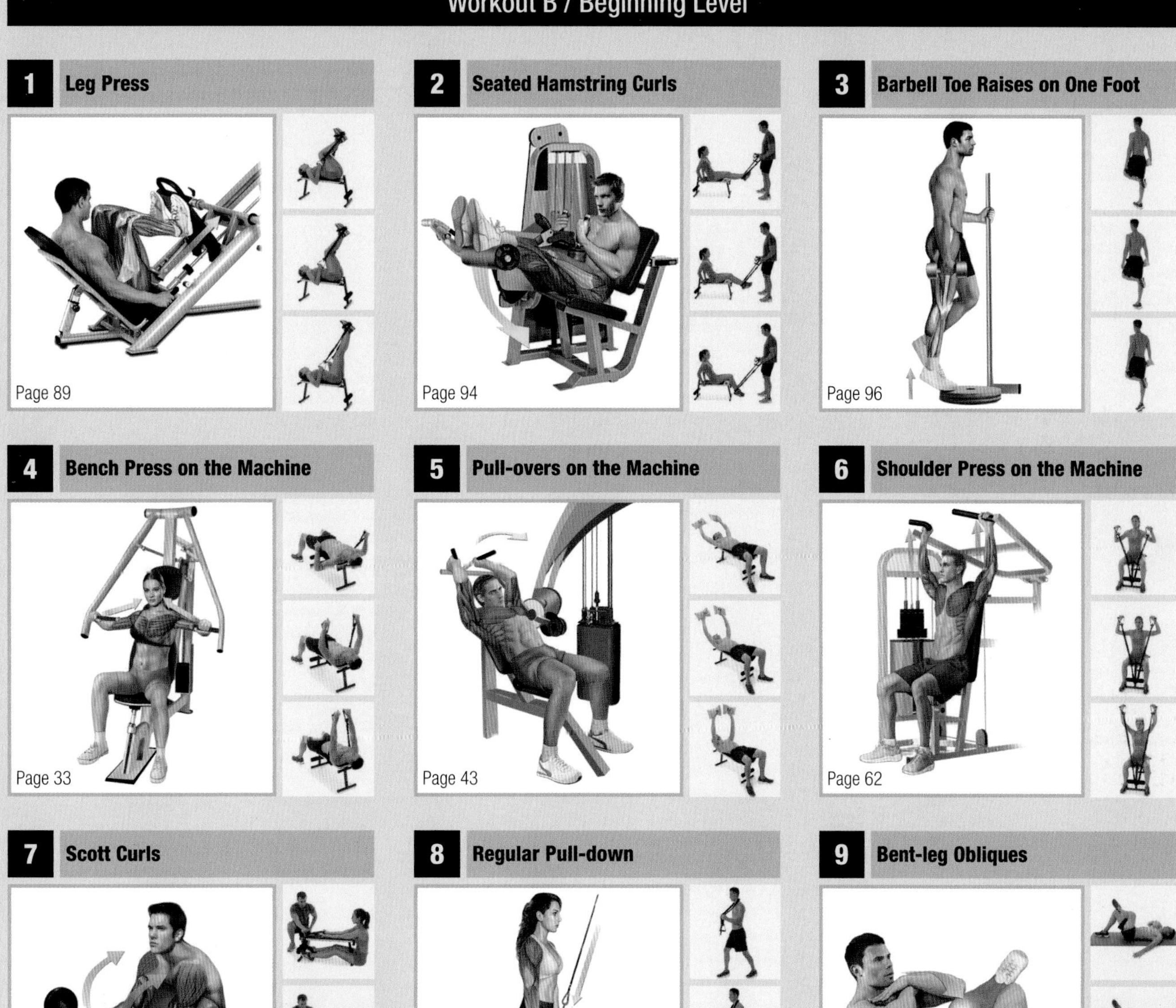

Training Level

Intermediate

Even if they are not experts, intermediate-level bodybuilders have mastered the basic techniques of the exercises with free weights, are thoroughly familiar with the sensation of muscle congestion, and can localize muscle work quite reliably. They already have muscles and structures that can withstand high-intensity work and allow them to work out with fairly high weights. For these bodybuilders, it is no problem to do exercises involving two joints or working at angles different from the ones used in exercises that are done standing or lying down on the floor (although these bodybuilders may experience a bit of instability in exercises involving a particularly complex technique).

The following workouts, among the many possibilities that could be chosen, are appropriate for intermediate bodybuilders, and they include exercises that can be done at the gym, as well as alternatives that can be done where there is no specific equipment. A bodybuilder may prefer certain exercises from Workout A and others from Workout B. Exercises can be exchanged from one workout to the other, as long as the substitutions are equivalent, so that exercises are done for

Workout A / Intermediate Level

1 Leg Press

Page 89

2 Seated Hamstring Curls

Page 94

3 Barbell Toe Raises on One Foot

Page 96

4 Cable Crossovers

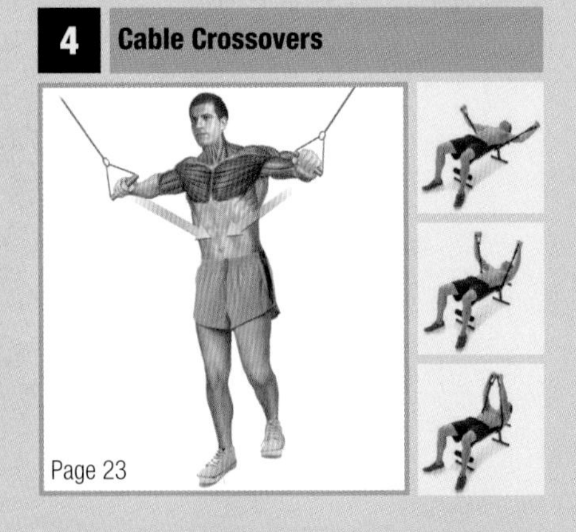

Page 23

5 Pull-downs with V-Grip

Page 45

6 Lateral Raises

Page 54

7 Incline Curls

Page 70

8 Seated Dumbbell Press

Page 76

9 Upper Body Cable Flex

Page 111

every muscle group and no muscle group escapes being worked. For example, a bodybuilder could opt to do Workout A, which starts with a Leg Press, but replace that exercise with the Hack Squat, which also addresses the quadriceps, but appears in Workout B. This way, bodybuilders can choose the exercises which work better for them, while preserving a complete and balanced workout. This book doesn't recommend putting together a workout based on exercises selected from different levels, because some advanced exercises may require a higher level of technical proficiency. Bodybuilders should remember that, if they are on this level, they should select the alternative material appropriately, so that the resistance level matches the one on which they are working. You must remember to follow the high-intensity training method described in the introduction.

Workout B / Intermediate Level

1 Hack Squat

2 Hamstring Curls

3 Barbell Toe Raises

Page 90

Page 93

Page 95

4 Bench Press with Barbell

Page 25

5 Reverse-grip Pull-down

Page 39

6 Seated Dumbbell Press

Page 61

7 Concentration Curls

Page 72

8 Ticeps Dips

Page 77

9 Upper Body Flex with Dumbbells

Page 112

Training Level

Advanced

Advanced-level athletes work out regularly and their results over time become obvious in their muscle growth. They can interpret perfectly the sensations of their bodies during training, and they have mastered the techniques of bodybuilding exercises, including ones that are complex and demanding.

Bodybuilders who are at this level are perfectly capable of designing their own workouts, including exercises that allow them maximum performance, discarding the ones that experience has shown have not contributed to achieving the intensity necessary for their muscle development.

Still, advanced bodybuilders can use the following workouts, because they are a good guide and an appropriate work tool for their level, with a spectrum of technically demanding exercises, as well as their variants for places that don't have bodybuilding equipment. Exercises that advanced bodybuilders

Workout A / Advanced Level

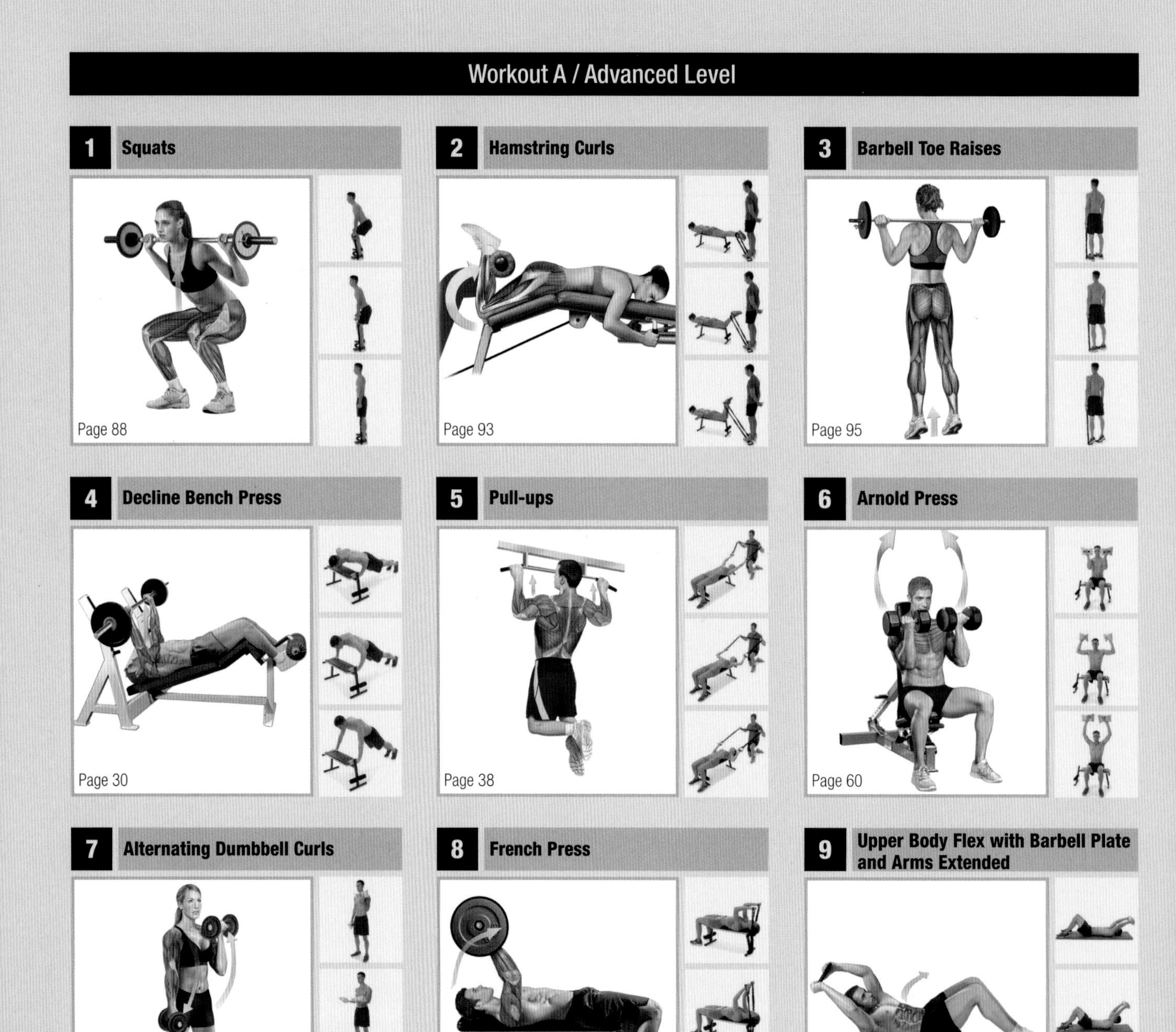

don't care for can be modified and replaced by others for the same muscle group, with which these bodybuilders feel more comfortable or with which they can get better results. Advanced bodybuilders know their strong and weak points, and they know how to work more of the latter in order to achieve a functionally balanced, aesthetically pleasing, well-proportioned body.

Once they reach this level, advanced bodybuilders have to be careful in selecting alternative material, because, in the case of stretcher bands, several resistance levels will be required. These bodybuilders will also have to handle moderate or heavy weights, which makes training without complete equipment increasingly complex.

Workout B / Advanced Level

1 Canadian Squats

Page 92

2 Seated Hamstring Curls

Page 94

3 Barbell Toe Raises on One Foot

Page 96

4 Flat Dumbbell Fly

Page 28

5 Inclined Row

Page 48

6 Military Press

Page 59

7 Standing Barbell Curls

Page 68

8 Triceps Back Kick

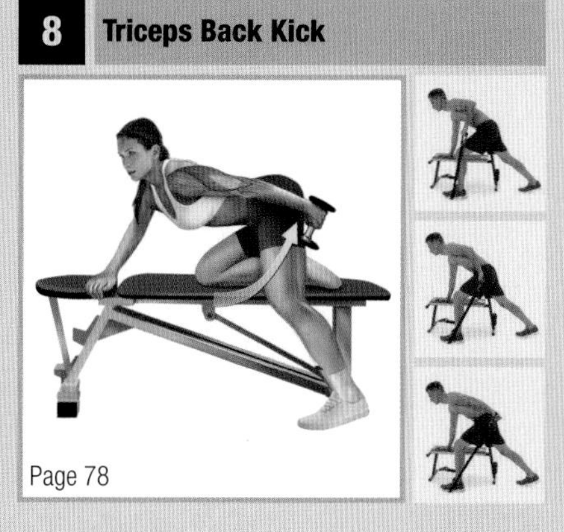

Page 78

9 Cable Crunches

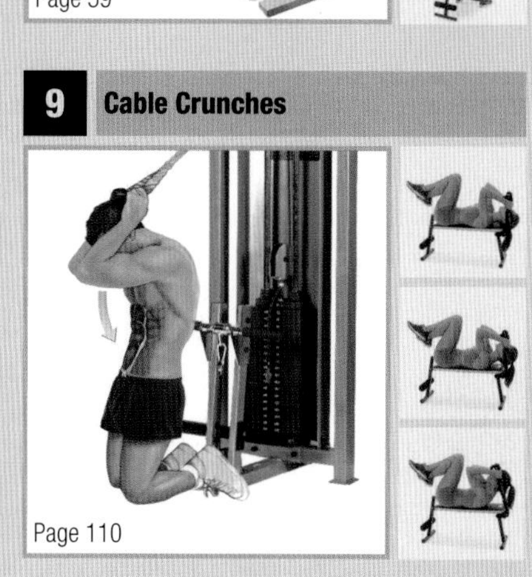

Page 110

Glossary

Abduction: a movement in which one moves a limb away from the central axis of the body. This distancing must be perceived from in front of the individual who performs it.

Adduction: a movement produced when one moves a limb closer to the central axis of the body. This movement must be perceived from in front of the individual who performs it.

Antepulsion: a movement equivalent to flexion, but applicable solely to the shoulder.

Apnea: an interruption of breathing that bodybuilders often experience at the instant of maximum exertion—for example, the holding of breath that takes place voluntarily during a very heavy bench press.

Aponeurosis: a type of tendon which has a flattened shape, instead of having the usual range of shape between cylindrical and conical.

Apophysis: a growth or projection on a part of the body. Usually, though not exclusively, this refers to prominent parts of the vertebrae (spiny and transverse apophyses).

Clean: a movement in which a weight lifter lifts the bar from the floor and places it on his or her shoulders.

Crunch: a movement that involves bending the upper body by contracting the abdominal muscles.

Diaphysis: an elongated, cylindrical part of a bone found between the two ends of long bones. This central part of the long bones tends to be longer and more slender than the epiphyses, or the ends.

Distal: the most distant area from the origin of a body part or from the center of the body. For example, the most distal part of the femur is the one closest to the knee and farthest from the hip.

Epiphysis: the end of a long bone. This is the area that forms a joint with other bones, and it tends to be broader than the diaphysis, or the central part.

Extension: a movement in which one pulls back a part of the body with respect to its central axis. This is the opposite in the case of the knee.

EZ-bar: a bar for bodybuilding with various bends incorporated into it, which places the wrist under less pressure in certain exercises, such as the French Press and the Barbell Curl.

Flexion: a movement in which one moves a part of the body forward with respect to its central axis. It is the opposite in the case of the knee.

Hack Press: a piece of equipment for the quadriceps and the gluteals in which the weight rests on the shoulders by means of cushions. The weight lifter moves up and down by means of flexion and extension of the hips and knees.

Insertion: the place where a muscle inserts. Normally, it is the distal end that is moved when a muscle contracts isotonically.

Jerk: a technical move in weight lifting where the bar is lifted overhead.

Knurling: a pattern or design on the metal surface of bars, grips for cables, and dumbbells, where the hands grasp them. Knurling provides a secure grip and keeps the hand from slipping, and it is an essential feature of working out with weights.

Kyphosis: physiological curvature exhibited by the spinal column in the dorsal region. The column curves toward the rear. This curvature is part of the proper makeup of the back as long it is not too pronounced. In the latter case, it is termed *hyperkyphosis,* and it is considered a medical problem.

Lordosis: physiological curvature in the spinal column in the cervical and lumbar regions. This curvature arches toward the anterior part of the body, and is a component of correct spinal conformation as long as it is not too pronounced. In the latter case, it is referred to as *hyperlordosis,* and it is considered a medical problem.

Multipower: a piece of equipment for working with weights, in which a bar is attached to a structure by means of guides that allow movement upward and downward, as well as holding the weight at different heights. The bar can be loaded with plates like a conventional bar, and, although the freedom of movement is more restricted, it offers greater security in certain exercises.

Negative Phase: the part of an exercise in which the resistance is greater than the force exerted against it, such that the muscle stretches and the weight moves by force of gravity. This is the instant where one lowers the weight. This is also known as the eccentric phase.

Origin: one of the insertions of a muscle, normally the proximal one, which does not move during the isotonic contraction of the muscle.

Physiological: referring to physiology, the science that studies the functioning of living beings and which is closely connected to anatomy.

Positive Phase: the part of an exercise in which the force applied overcomes the resistance, and, thus, the muscle shortens and the weight moves opposite to the force of gravity. This is the instant in which one raises the weight. This is also known as the concentric phase.

Press: an exercise in which the weight is pushed, such as the bench press, the French press, and the military press.

Pronation: a rotational movement in the forearm that makes it possible to put the back of the hand upward and the palm downward.

Proximal: an area located closer to the origin of a body part or closer to the center of the body. For example, the proximal part of the femur is the one closest to the hip and farthest from the knee.

Pull-over: an exercise in which the weight is moved downward, whether on a bar or a dumbbell, behind the head, starting in a prone position on the back. This exercise can focus on the pectorals or the back muscles, based on the width of the grip.

Retropulsion: a movement equivalent to extension, but applicable solely to the shoulder.

Rotation, External: a movement in which one moves a part of the body away from its central axis within the transverse plane.

Rotation, Internal: a movement in which one moves a part of the body toward its central axis within the transverse plane.

Scott Bench: a type of bench for doing biceps exercises in which the bodybuilder sits and rests his or her arms and armpits on a front cushion. This bench gets its name from bodybuilder Larry Scott, who popularized it in the 1960s, even though he did not invent it.

Supination: a rotational movement of the forearm in which it is possible to place the palms of the hands facing upward.

Tuberosity: a protuberance that exists at the end of some bones and that commonly coincides with the insertion of a muscle.

Bibliography

Ashwell, Ken, *The Anatomy of Stretching* (Hauppauge, NY: Barron's Educational Series, Inc., 2014)

Ashwell, Ken, *The Student's Anatomy of Exercise Manual* (Hauppauge, NY: Barron's Educational Series, Inc., 2014)

Baechle, Thomas R. and Roger W. Earle, *Essentials of Strength Training and Conditioning,* 3rd Edition (Champaign, IL: Human Kinetices, 2008)

Contreras, Bret, *Bodyweight Strength Training Anatomy* (Champaign, IL: Human Kinetics, 2013)

Delavier, Frederic, *Delavier's Core Training Anatomy* (Champaign, IL: Human Kinetics, 2011)

Delavier, Frederic, *Strength Training Anatomy,* 3rd Edition (Champaign, IL: Human Kinetics, 2010)

Evan, Nicholas, *Bodybuilding Anatomy* (Champaign, IL: Human Kinetics, 2006)

National Strength and Conditioning Association (NSCA), *Exercise Technique Manual for Resistance Training,* 2nd Edition (Champaign, IL: Human Kinetics, 2008)